T0358329

Ortega's "The Revolt of the Masses" and the Triumph of the New Man

Ortega's "The Revolt of the Masses" and the Triumph of the New Man

Pedro Blas Gonzalez

Algora Publishing
New York

ISBN-13: 978-0-87586-470-9 (trade paper)
ISBN-13: 978-0-87586-471-6 (hard cover)
ISBN-13: 978-0-87586-472-3 (ebook)

Library of Congress Cataloging-in-Publication Data —

Gonzalez, Pedro Blas, 1964-
Ortega's The revolt of the masses and the triumph of the new man / Pedro Blas Gonzalez.
p. cm.
Includes bibliographical references (p. 183) and index.
ISBN 978-0-87586-470-9 (trade paper: alk. paper) — ISBN 978-0-87586-471-6 (hard cover: alk. paper) — ISBN 978-0-87586-472-3 (ebook) 1. Ortega y Gasset, Josi, 1883-1955. Rebelisn de las masas. 2. Ortega y Gasset, Josi, 1883-1955—Political and social views. 3. Civilization. 4. Proletariat. 5. Europe—Civilization. I. Title.

CB103.G66 2007
901—dc22

2006102139

Front Cover: Hand and Face, Creator: Jim Dandy, ©Corbis

Printed in the United States

"There are no longer protagonists;
there is only the chorus."

— José Ortega y Gasset
The Revolt of the Masses

To my wife, Anne, in love and appreciation. And to my children, Isabella Sophia and Marcus Julian, for the joy you have given me.

Table of Contents

By Way of an Introduction

T. S. Eliot begins his introduction to Josef Pieper's seminal work *Leisure: The Basis of Culture* by invoking the contemporary state of philosophy. Eliot, who is hardly a new comer to the discipline, frames the question in a manner that takes into consideration technical matters and where this venerable discipline found itself during the middle of the twentieth century. But more importantly, Eliot addresses the fundamental question of temperament and philosophical vocation. He looks to an ideal time when the day will dawn again when a philosopher will come forth "whose writings, lectures and personality will arouse the imagination." But even more relevant to our present condition, Eliot explains, is that philosophy must begin again to exercise its former, more meaningful etymology — "the need for new authority to express insight and wisdom."

It was not many years later that several such figures would begin to make headway in at least some of Eliot's prescribed categories: Sartre, Camus, Marcel and Jaspers come to mind as embodying aspects of the aforementioned philosophic qualities. With the notable exception of Sartre, history has vindicated these other figures for their insight and wisdom.

Ostensibly, Eliot goes on to say that at the end of any philosophical process what remains — in fact, what allows for insight and wisdom — is what makes philosophy indispensable to reality: common sense.

Thus, a restoration of the philosophical discipline must contain enough respect for the dignity of man — individual subjects — to garner other possible alternatives beyond the currently destructive "biological entity" and the unprecedented surge in anti-humanism. A fine start to the restoration of philosophy as well as the humanities is a renewed concept of man as an end in itself. Man cannot continue to be subservient to utility. Pieper cites the august Goethe: "I have never bothered or asked in what way I was useful to society as a whole; I contented myself with expressing what I recognized as good and true. That has certainly been useful in a wide circle; but that was not the aim; it was the necessary result."[1]

When Pieper writes, "leisure, it must be remembered, is not a Sunday afternoon idyll, but the preserve of freedom, of education and culture, and of that undiminished humanity which views the world as a whole." This recalls José Ortega y Gasset's exaltation of the self in his description of the aforementioned in *Meditations On Quixote*. In that work the reader is summoned to listen to the "pounding of his own heart."[2] Here the emphasis is on reflective silence. What Ortega sets out to describe in *Meditations On Quixote* is nothing short of a phenomenological analysis of the self. This task, however, like most phenomenological/existential approaches to life, proved to be short-circuited by the very weight of the words used to describe it.

1. Josef Pieper. *Leisure: The Basis of Culture*. Indianapolis: Liberty Fund, 1999, p. 33.

2. José Ortega y Gasset. Meditations *on* Quixote. New York: W.W. Norton, 1963, p. 57.

Initially, we are shocked to notice the silence of the forest. Yet what we notice is not so much silence but the absence of sound. Reality, in this sense, is encountered as a negation of the ever-present bustle, the clamorous daily world of man. This absence of environmental impetus forces us to experience not an absolute silence, for Ortega argues that this can never be achieved, but rather a form of silence that directs a reflective gaze to itself.

What initially seems like an occasion for reflection, in many cases, Ortega goes on to argue, becomes an uneasy, even an existentially heavy burden. When the external noise gives way to such a surprising and challenging silence, "all this is disturbing because it has too concrete a meaning."[3] Yet this concreteness is not encompassed by "theory." It is instead a lived vitality. What is encountered in this silence is the fragile and radical strain of the self stripped of all societal trappings. This is not man as *homo faber*, but rather as an entity that does not readily know how to react to this naked existence — and who consequently knows not what to do.

A possible antidote Ortega suggests for this often frightening experience is to bargain for a form of silence that is "purely decorative," where "unidentifiable sounds are heard."[4] To fill this silence we must cross back into the clamor and noise of preoccupation with the social, that is, with external reality — we must lose ourselves in things. This might be regrettable, Ortega argues, but this signifies the common way of life for most people.

Meditations on Quixote achieves precisely what it sets out to achieve: a meditation. The book is a meditation on the nature of human reality and how this is appropriated by man. But if Ortega argues that man is a social entity, he also establishes the conditions for this social interchange to take place. What is important, however, in this social friction, if not fracture, is equanimity. The corresponding pole of man's social (or what amounts to his external) condition is

3. Ibid., p. 57.
4. Ibid., p. 58.

garnered by the interiority that he recognizes in himself. It is the latter that is encountered when external worldly clamor is refused its sensual stranglehold on man. What is gained instead, after the temptation of popular noise has been effaced, is nothing less than existential human autonomy.

Ortega explains autonomy as the possession of an inward sense of life. This inward turn where the seemingly biological and external public persona becomes self-aware is the true starting point of all philosophical activity. Yet philosophy, Ortega is quick to point out, is not just a process that seeks to uncover the profound, but also one that is equally concerned with the spurious or superficial aspects of the human condition. When he refers to seeing he does suggest that it is merely a sensorial function. The eye only "intends" and in doing so it removes the object out of what is up to that point an undistinguishable multiplicity. Thus to casually glance over things negates the inward, three-dimensional quality of reality. But equally damaging to the inherent structure of reality is its careless dismembering, where what is left is a vacuous transparency. Ortega explains:

> And if we succeed in obtaining layers so thin that our eyes can see through them, then we do not see either the depth or the surface, but a perfect transparency, or what is the same thing, nothing. For just as depth needs a surface beneath which to be concealed, the surface or outer cover, in order to be that, needs something over which to spread, to cover.[5]

Ortega, like Pieper, recognizes the objectifying nature of work. The concern is not with the value of work itself, because this much Ortega views as a positive having-to-do that safeguards most people from the devastating effects of idleness and boredom. Instead, the objectifying aspect of work has to do with its ability to remove us from ourselves, as it were. He makes this clear in *The Modern Theme*, where he views modernity as a form of ushering man out of himself and displacing his vital grace with artificiality. The toil of work is

5. Ibid., p. 62.

countered instead by the notion of sport, an Ortegan notion that is closer to a form of reflection than it is to mere play. The sporting attitude, he tells us, is a morally heroic stance toward reality. Hence the equation "work is to external reality as reflection is to the interior life as radical reality" is perhaps nowhere more evident than in *Meditations on Hunting.*

Meditations on Hunting exhibits that supremely stealthy characteristic found in all of Ortega's work: philosophical profundity achieved through an exemplary clarity. This work is as much about hunting as *Meditations on Quixote* is about the concern of Cervantes' spirited Don Quixote with the nature of truth. Instead of compounding philosophical thought with portentous titles and layers of self-referential jargon, his work beckons the conscientious reader to reflect in a noble way that rejects the "inertial thinking" of philosophical categories. And instead of showcasing an old, tired and trite scholasticism, his thought weaves through the history of philosophy as if they were merely a series of signposts on the path to reflective thought. But scholasticism, he informs us, is the opposite of philosophy — a truism that remains vital — regardless of the animated protestations of self-preservation issued from contemporary critics and "theorists." Instead, the essence of originality is that it does not purport to call attention to itself. As an example of this, we have Ortega's notion of death in *Meditations on Hunting*, when we discuss the triumph of the new man, especially in "postmodern" philosophy:

> But this is precisely what death is. The cadaver is flesh which has lost its intimacy, flesh whose "interior" has escaped like a bird from a cage, a piece of pure matter in which there is no longer somebody hidden.[6]

Equally important to Ortega's description of man's self-discovery in the forest is the realization of the inward quality of human existence. Louis Lavelle's *The Dilemma of Narcissus* is an enchanting phil-

6. *Meditations on Hunting*. Translated by Howard B. Wescott. New York: Charles Scribner's Sons, 1985, p. 91.

osophical study on the nature of the self. Lavelle reminds us that Narcissus' "own beauty has become a tormenting longing, which separates him from himself by showing him his image, and which drives him to seek himself where alone he sees himself — namely, where he has ceased to be."[7]

In addition, we can compare Lavelle's statement citing Narcissus' emptiness: "But Narcissus cannot bear either to be or to act: as that subtle man Gongora puts it, he is reduced to 'calling forth echoes while discovering their origin' " with Ortega's mass man who is incapable of leadership but who refuses that others do so. Thus, human autonomy ought not to be confused with what some consider a vague and lazy individualism. Apparently, the critics of individualism have erroneously collapsed the two concepts. Narcissus' problem is that he is essentially torn between two semblances of himself: the figure reflected in the water and the one who stares into it. Both are equally vacuous entities. The figure that Narcissus witnesses in the water is not recognizable as his self — rather as "himself" — or what practically amounts to another figure. Lavelle adds: "If Narcissus went down to destruction, it was because he actually tried to create this duality in his very being. For he thought he could see himself and enjoy himself before he acted and before he had made himself."[8] Hence, what Narcissus lacks is sincerity. And sincerity is nothing less than "the attention that arouses our potentialities."[9]

Not unlike Narcissus, Ortega's trek through the forest also demands a degree of sincerity in the act of truth finding. The forest is to Ortega what water is to Lavelle's Narcissus: a confrontation with appearance. The silence that is encountered in the forest leaves few avenues open for distraction. Instead, its effect is felt in providing an ample mirror for self-reflection.

7. Lavelle, Louis. *The Dilemma of Narcissus.* Translated by W.T. Gairdner. Burdett, New York: Larson Publications, 1993, p.33

8. Ibid., p. 70.

9. Ibid., p. 71.

Regardless of how much Narcissus loses himself in his regard for his outward image — what amounts to his body — he cannot help (both Lavelle and Ortega suggest) noticing that his body is part of a greater circumstance. While in the forest Ortega, too, cannot help but to reach a stage of existential awakening where he becomes "I and my circumstances." The circumstance part of this equation is nothing less than my life, or all that happens to me, but it is not "I" properly speaking. I am bound to the world through my circumstances. Lavelle refers to this as sensibility. He explains: "The individual's sensibility joins him to the All, and yet the distinction between them is not abolished."[10]

Hence as we delve ever deeper in Ortega's thought we can arrive at the understanding that the phenomenological and existential themes contained in his work more often than not exist as latent possibilities for man. And much like these themes, the chosen manner utilized to communicate them is equally indirect at times. Ortega explicitly says in the first section of *Meditaciones Del Quixote* that the totality that is the forest exists as vital/existential possibility.[11] But what ought we to make of Ortega's notion of possibility? He quickly proceeds to answer this question in the following section, titled "Profundidad y superficie" (Profundity and Superficiality), where he explains that the purpose inherent in patent reality is to direct our attention to what is latent.

Thus the importance of possibility in Ortega's work does not strictly negate the level of the patent in favor of a "profounder" understanding of reality. At least he does not go as far as to violate the relevance of the seemingly superficial aspect of human life. This means that these two poles must be construed and accepted in

10. Ibid., p. 94.

11. See: Julian Marias. *Metaphysical Anthropology: The Empirical Structure of human Life*. Translated by Francis M. Lopez-Morillas. University Park: Pennsylvania State University Press, 1971. "Philosophy is present only if man believes that he can progress from the patent to the latent, to uncover it and account for it; but this is possible only if the *real* has *consistency*. In taking progressive possession – sometimes regressive possession, we must not forget — of this connection the first rational interpretation of reality as nature or (physis) has occurred," p. 9.

equal measure as a duality of a single whole. The demarcation point between the phenomenal and the noumenal is the privacy that human existence represents for itself. But human existence is not merely construed as biological life, rather more on the lines of "this particular life that runs through me." Seen as such, we can view the purpose of human existence as consisting of the need for synthesis. The latter, too, Ortega warns us, is a form of synthesis, even in this embryonic and superficial self-negating mode.

The realization that the task of human existence is to seek perpetual synthesis is an essential component of Ortega's work. From this we are better able to understand life as lived immediacy and latent possibility is complimentary to reflective life. This ordeal is no less than Ortega's delicate balancing of life as vital existence and the life of the intellect. The key here, he explains throughout his collected work, is not to crush the spontaneity of the former with the calculating nature of the latter. This, then, is an accurate portrayal of the philosophical vocation par excellence. This is also Ortega's general approach to the history of philosophy and its inherent exigencies.

Philosophy in the First Half of the Twentieth Century

When in 1951 a Brazilian philosophical journal asked José Ortega y Gasset (1883–1955) to write a series of articles concerning the then current state of philosophical affairs during the first fifty years of the 20th century, Ortega was already 68 years of age and had been suffering from stomach cancer since his 50s. Also, by this time he had enjoyed a degree of fame, rare in philosophical circles, which was not restricted to Spain and Latin America but that also included the United States and all of Western Europe. For instance, he was highly respected as a writer and thinker by Thomas Mann, and was referred to by Albert Camus as being "after Nietzsche, perhaps the greatest European writer."

Ortega was a very visible figure. During the early part of the century and leading up to the events of the Spanish Civil War he had a great deal to say; and the most accessible media that he could employ, at the time, to accomplish this task was the newspaper. His many and often diverse newspaper articles brought him considerable attention, both in philosophical circles and especially from the public at large. Whether this attention came in the manner of criticism or approval is irrelevant, for while today academics are largely critical, through his many articles he found an audience that would furnish him with the necessary feedback to test his most pressing convictions. In this respect, Ortega can also be credited with attempting to take the level of the journalism of his day to a more sensitive and profound awareness of intellectual concerns. He felt that newspapers should not merely be concerned with reporting events.

For this reason in 1923 he founded a literary and philosophical journal titled *Revista de Occidente*, which, even though it initially enjoyed only a short run, nevertheless received some degree of critical acclaim. In 1948, in keeping with his humanistic concerns, Ortega, along with the Spanish philosopher and disciple Julian Marías, founded the "Instituto de Humanidades" in Madrid.

The offer to write a short history of philosophy must have seemed to him like a timely opportunity to make his views on historical reason heard once again; especially given the fact that by that time he had already slipped from the forefront of philosophical thought in Spain, a position which he had held up to about the 1930s. After all, Ortega is, quite arguably, Spain's best-known thinker after Miguel de Unamuno. He was a philosopher in search of the vital pole of the human condition. This is readily seen if one realizes that his newspaper articles were intended to bring the sphere of philosophical discourse to a general readership. Reflection, contemplation, and overall existential concerns were, in his estimation, the sole business and backbone of any genuine attempt at philosophizing. The problem, then, as he saw it, manifested itself in the form that: if philosophical reflection is the natural call of the thinker *per se*, vital, spontaneous life is the concern of all alike. The synthesis of these two seemingly odd poles was the inspiration of his life's work.

The article, even though never completed, works well on two levels. As a brief history of philosophical thought during the first half of the twentieth century, it serves as an important eyewitness to Ortega's intellectual milieu. As he states in the article, "this chronological trajectory coincides with my life; indeed, no one can tell me this history given that it is my own existence." But this "eyewitness account" can be rather misleading at first, given that Ortega does not begin his inquiry at the beginning of the twentieth century, but rather, in true Ortegan fashion, he proves to be evasive, often deviating from the point at hand, pausing to interrupt himself with afterthoughts and asides, including a description of Hermann Cohen that was not entirely becoming to a thinker of his stature. The essay, in fact, begins not at the start of the twentieth century but rather with the ancient Greek philosopher Parmenides.

Ortega begins the article — keep in mind that this was to be the first of several articles — by attempting to tackle the ancient Greek notion of being. While it remains true that the task that he had before him was a formidable one, nevertheless from the very start it conveys something of Ortega's own inherent philosophical preoccupations. We may consider them in this case in particular, I believe, to be the opportunity to make known his views on historical reason once again along with one more opportunity to show his aversion to neo-Kantianism.

Hence, the reader is to embark along with Ortega on a historical trip, all the way back to the very beginning of philosophy, with the Greek treatment of the newly discovered problem of being. Ortega's apparent overall intention in proceeding in this manner is to attempt to depict how it can be that nothing substantial, in his estimation, has in fact been added to philosophical thought by modem philosophers. It is Ortega's contention, in fact, that in order to actually undertake any real and genuine philosophical activity today, the entire notion of being must be abandoned. This may sound rather paradoxical given that he begins his inquiry with Parmenides. But he contends that continuing a self-conscious harassment of the question of being is a waste of time given its direct and pre-calculated manner. Instead, his approach takes into consideration the prior attempts to address this problem; and he therefore arrives at the conclusion that it is best to allow this problematic to yield its own truth. This is what he understands the Greeks to have accomplished, seemingly effortlessly. This, then, is the general impression with which the article leaves the reader.

But this is precisely what Ortega has attempted to do throughout his work: he treats thought as a necessary evil at the very service and core of life itself. In other words, he tried to fuse a self-conscious and rational appeal to reflection with the spontaneous and overall unpredictability of vital life. His reasoning being, that after the legitimate and spontaneous treatment that the ancient Greeks gave to the notion of being, to attempt any further efforts is truly to waste one's time. In this respect he criticizes Heidegger, whom he otherwise thinks very highly of, for engaging in an unfruitful and stubborn academic exercise.

Also, in a still more subjective vein, the article serves to vent Ortega's overwhelming distaste for the new materialistic direction that philosophical thought had taken. For this reason he goes back to the ancient Greeks and their concern with Being. And what, then, was this new philosophical direction, as far as Ortega was concerned? The problem for him, as countless other thinkers have also observed, was that modem philosophy in general made the grave mistake of attempting to emulate the mathematical sciences. In other words, philosophy according to Ortega had practically abandoned its call to confront reality systematically by utilizing the only tool it possesses, reason. By reason Ortega means vital reason, of course; or the

Heraclitian condition whereby using "life" itself as an organism for knowledge, truth can begin to subtlety expose itself.

What Ortega argues for in this brief history of philosophy is something that he has otherwise made explicit throughout his work, mainly his conviction that strictly speaking philosophy as an activity or manner of thinking that faces naked reality, holistically, ended long ago with the ancient Greeks. All subsequent philosophical endeavors have been merely a rehashing or, simply articulated, an academic commentary on the pre-existing philosophical canon. This latter activity he saw as pertaining to the history of philosophy, but he did not regard it as philosophy. Philosophy, as a vital and life-forging way of life, he argued, had played out its originality (and thus had run its course) long ago.

Ortega instead viewed post-Cartesian philosophy as turning away from itself as a method of confronting the enigma of being and thus embracing the realm of the mathematical sciences. In this respect most of modem philosophy is characterized by Ortega both as a pseudo-philosophy and pseudo-science. This was not philosophical activity in his estimation. What looked like philosophy was in reality merely a decaying shell; his main contention here being that philosophy in the modern world had taken up a new and inane direction.

Moreover, at the core of this critique of modernity is Ortega's insistence that as ancient Greek philosophy unfolded, it served its ultimate role as a way of life. In this respect Ortega went head to head with the many philosophical movements of the early part of the twentieth century that attempted to establish philosophy as a rigorous or pure science. Philosophy for Ortega cannot be pigeonholed as just an academic or professional activity; it must serve as a tool for life itself, a life process, a *Weltanschauung.* Ancient philosophy, it can be argued, can in fact be seen as a catharsis whereby the inquietude and anxiety of human life can be exorcised, if not altogether controlled. The process of philosophizing, Ortega would argue, is one of a moving away from a mytho-poetic cosmology and into a rationalistic manner of viewing the universe; and in this regard it was a process of engendering rationalism. But this cannot be oversimplified, because just as mytho-poetic cosmology served as an explanation for the cosmos, so too the developing rationalism was at the service of conscious human life. In other words, Ortega saw the strength of philosophy as resting on the idea that it is a way of life and not just a scientific technique or method.

This point then brings us to the activity and role of the philosopher. For Ortega the philosopher is one who embraces the dictate of a "rigorous manner of thinking," one which challenges the thinker to investigate the nature of being. This vital organon, once again, seems to run diametrically against the grain of the method of the modem sciences, in his view, whose method consists in being a theory of knowledge and which neither claims to know nor seeks to know the ultimate nature of being. The method of modern science, Ortega emphasizes, is only concerned with epistemology, while that of philosophy has always been an ontological one. In Ortega's work, this Aristotelian organon is the vital phenomenon that he regarded as life as radical reality.

Radical reality is to be viewed as being the original condition of primordial man. But also, in a greater sense, this notion of radicalness is conveyed by the facticity of every instance of human existence. Thus Ortega's central critique and profound distaste for the method of the physical sciences and the alluding pretensions of modern philosophy is that, on one hand, their method is disconnected by the nature of the diverse fields of inquiry that have resulted after the growth and subsequent spreading of the scientific method. This is coupled with the nature of inquiry of the physical sciences, which is essentially concerned with questioning not the "why" of things, but rather the "how." This Ortega sees as a serious deficiency of the sciences, in that he views it as a turning away from the immanent problems of life itself, which can only be solved by introspection and contemplation.

On the other hand, Ortega holds that the scope of philosophy should always be synoptic. Philosophy, then, is that human method of engaging reality on its terms. The philosopher is the thinker *par excellence.* Hence there is a sense in which Ortega engages the reader of his time in a line of questioning that says, "tell me the immediate concerns of philosophers during any given age, and I will tell you their philosophies." For this reason he stated many times throughout his work, including in this essay, that, "autobiography is the superlative of historical reason." But what can autobiography ever have to do with philosophical thought, some may ask? To Ortega this was not a relative or arbitrary union, but it was rather at the center of what it meant to philosophize. It remained true that philosophy was a rational endeavor; this he would not refute. But "rational" also meant for him to be vital in scope. In this respect philosophy was always to remain a subjective endeavor.

This brings us to the import of his notion of historical reason. What then does Ortega mean by historical reason, or vital reason, as he has referred to it? Vital reason serves as nothing less than the core of all subsequent forms of reason, given that all epistemology depends upon the subjective reason. Vital reason is then a self-conscious preoccupation with one's very life; one's being. Vital reason serves as a Protagorean impulse to view man as the center of the universe. This would be the case for Ortega because all knowledge of the universe must first proceed from an inherent awareness of one's personal circumstance. It is in this respect that Ortega's notion of "I am I plus my circumstances," which is a direct offshoot of his concept of radical reality, can also be seen as philosophical humanism. For Ortega the autonomy of the self firmly rests on the principle that man is in constant strife with human reality and therefore to know this is to begin to seek authenticity. This was clearly not the case in twentieth century philosophy, as one can gather from this article by one of the greatest Spanish philosophers of all time and a man who was also a witness to the truth of the matter of which he so eloquently writes. Nowhere is this perhaps better articulated than in his subtle yet incisive, *The Revolt of the Masses.*

Chapter 1. Revisiting The Revolt of the Masses

The development of mass man sees its culmination in what Ortega considers to be an invertebrate history. To become well acquainted with *The Revolt of the Masses* it is first necessary to understand its genesis. *Invertebrate Spain* (*España Invertebrada*) is essentially a treatise on history, specifically some aspects of Spanish history. But it is not until the second part of this work, "La Ausencia de los Mejores" ("*The Absence of the Best*"), that Ortega develops his notion of mass and noble man. He writes:

> In essence: the absence of the best has created in the masses, in the "people," a secular blindness that cannot distinguish between the best and worst...[12]

The history of mass man results in the overturn of the rational logos from the moment that it begins to view itself as a solid, or what amounts to a sovereign, historical movement. While *Invertebrate Spain* takes a particular, seemingly isolated subject matter — Spain — as its starting point, one ought not to be fooled by a false provincialism. *Invertebrate Spain* is a study of the metaphysics of the rigor

12. Ortega y Gasset, José. *España Invertebrada: Bosquejos de Algunos Pensamientos Históricos*. Madrid: Revista de Occidente, p. 154 [my translation].

and care of existential autonomy that often goes unnoticed by the casual reader. Ortega cites a good example of this when he writes:

> But the historical norm, which is met in the case of Spain, is that notions degenerate due to intimate defects. Whether man or nation, their vital destiny is dependent on its radical sentiments and character.[13]

Ortega's preoccupation in this work is not a sociological one. In the course of writing there will inevitably be instances of citations where statistics, historical models and demographical data will be utilized. Plato, too, made use of empirical references such as the human eye and the Sun to depict a universal notion of truth. But what is particularly significant about Ortega's project in *Invertebrate Spain* is that his analysis of man and culture perhaps can be equally applied to other disciplines at the turn of the century while only substituting the particular. But how can that be the case, some critics will object? The immediate answer to this question has to do with the very act of philosophizing. While it remains true that Ortega's task is one of a particular cultural historical hue — a thinker in a given epoch — this does not invalidate the rational logos at work. What Ortega sees at stake here is nothing less than a perpetual struggle against the aberrant tolerance of moral/intellectual and spiritual mediocrity. What was to be the fate of the newly liberated slave in Plato's cave? To return and enlighten the others, perhaps? And who is said to carry the burden of truth within him? Certainly not those who remain in the cave.

The last four pages of *Invertebrate Spain* demonstrate at least two significant things: 1) The universality of Ortega's project of distinguishing the noble from mass man, the former of these concerns, preoccupied both Socrates and Plato as well. Ortega views this daunting task with cautious but restrained optimism when he writes: "...the publication of these pages will result to be pointless, but harmless: They will neither be understood nor paid attention

13. Ibid., p. 159 [my translation].

to."[14] From there he goes on to offer a possible solution to the problem of mass man: "The mission of the masses ought to be nothing other than to follow the lead of the best, not to supplant them."[15] The other telling aspect of *Invertebrate Spain* as a work of philosophy is its projection of a future work that will continue his analysis. That work was to surface as *The Revolt of the Masses* the following year.

And Just Who Is the New Man?

Ortega concludes *Invertebrate Spain* by posing the question of what will become of the masses in Spain? Will they throw off their invertebrate rebellion and accept the responsibility of "confronting the most vulgar situations of existence?"[16] Or will mass man continue to view itself as the grindstone of an inane politics? Undoubtedly, the "revolt of the masses" is a term that does not mean *prima facie* what some might think. The phrase does not signal entire herds of people "revolting against some genuine or alleged evil," even though examples of this can be found in recent times. Nor does it depict individual rebels without a prescriptive cost. What the phrase does suggest — as Ortega points out — is a historical turning on its head of all valuation. What is the reason for this? The answer, surprisingly, in many instances is nothing more complicated or sophisticated than existential discontentment and boredom, the promise of a "liberating" nihilism, or the negation of responsibility for individual destiny.

It can be argued that the twentieth century was the century of the masses, especially the rupture with traditional modes of life in the latter half of that century. But we must make clear that "mass" does not mean mass culture, mass hysteria, mass production, mass transportation or even masses of people. These are all phrases that

14. Ibid., p. 162 [my translation].
15. Ibid., p. 162 [my translation].
16. Ibid., p. 163 [my translation].

convey quantity. "Mass," "the masses" and "mass man" are instead understood here as descriptions of quality.

While the French Revolution is indicative of a historical moment of mob rule, the totalitarian systems of the twentieth century have used the term "mass" to indicate a mob mentality that can easily be molded for any suitable end. What mob rule was to the French Revolution the new man is to the newly forged behaviorism of communism and socialism. The very essence of the new man is that of an existential zombie who must not be allowed to retain any semblance of an autonomous will. This new man — *homo novus* — an entity that possesses no reality for itself and who makes nothing of its own existential condition, is made an agent of political power by negating and slashing axiological hierarchies. Roger Scruton writes in *A Dictionary of Political Thought*:

> Since man has a "historical essence" he is, in one sense, not the same under the new economic order, and values and aspirations that previously motivated him may no longer be understood or recognized.[17]

The new man, then, is comprised of several distinct but equally centralized characteristics. Contemporary man is said to inhabit a technological world. Technology serves as a major component in the life of the new man. The technological temptation serves as a perfect fit for the new man because it serves to alleviate or assuage the "burden" of life. Technology is to the new man what forced agricultural and military service were to twentieth-century totalitarian societies.

Another dominant present-day aspect of this entity is a rampant de-sensitizing of this new form of consciousness. This new con-

17. Scruton, Roger. *A Dictionary of Political Thought*. New York: Harper & Row, Publishers, 1982. The historical significant of the new man bears pointing out. Thus, the reminder of his entry seems warranted: "The phrase 'new man', 'new Communist man' or 'new socialist man', has bee used since the 1920s by both supporters and critics of USSR Communism, in or order to describe the transformation not only of the economic order but also of the individual personality, that has taken place or should take place, either under socialism, or under the 'full' communism to which socialism supposedly leads," p. 322.

sciousness pays next to no attention to existential and metaphysical concerns. These are viewed as a waste of time, given the new preoccupation with the external world. While critics and supporters of this de-sublimation alike have concentrated disproportionate levels of attention on this concern, a deeper understanding of this phenomenon has evaded them. With increased and totalizing disregard for the sublime in human existence also comes a de-sensitizing of aesthetic sensibility — a prerequisite condition — for what Unamuno has called the "tragic sense of life."[18] This second component of the new man is essential for the attainment of a well balanced and vitally felt aesthetics of life. The major implication in this respect is not that the notion of God is eroded, but rather that the possibility of viewing man as a cosmic phenomenon, in its varied forms, also vanishes from such a new form of consciousness.

Following from this significant psychological paradigm shift is the sense of rootlessness that accompanies the lack and need for transcendence. Human concerns with the seemingly dualistic nature of being / becoming, finite / infinite, and immanence / transcendence, to mention just a few everyday metaphysical quiddities that cannot be permanently deleted from human consciousness.

Another major feature of this new man that Ortega makes known is the politicization of reality. This is perhaps the most devastating to a genuine humanism. By subjugating vital life to the yoke of politicization, the new man begins to feel omnipotent over nature. Perhaps Auguste Comte was correct in assuming in *Cours de philosophie positive* that modernity would eventually turn all its energy into a pseudo science of man as our main earthly preoccupation. Hence the pulse of the new man is being taken with the instruments of positivism: radical skepticism, physicalism and moral and spiritual cynicism.

The Revolt of the Masses is such an insightful work precisely because the analysis that Ortega offers is a universal prescription of the com-

18. Miguel de Unamuno. *Del Sentimiento Trágico de la Vida*. Madrid: Editorial Plenitud, 1966.

ing of a full-blown positivistic age. While Ortega does not argue for such an age as the definitive "development" of man, like Comte he does offer a substantive explanation of its causes and overall direction. *The Revolt of the Masses*, then, is a narrowing refinement of the themes that its author began to formulate in the second part of *España Invertebrada*. Ortega begins his analysis of mass man with an appropriately titled chapter: "The Coming of the Masses." While Ortega recognizes the coming of the masses to power as an historical fact, however, nowhere does he suggest this to be merely a historical phenomenon. From the outset Ortega refuses to make his analysis a social/political diatribe. The words "rebellion," "masses," and "social power," he tells us, are not to be construed as being "exclusively or primarily political."[19] This qualification is of profound importance given that at the time of writing Soviet rule had already been established in Moscow since October 25–November 2, 1917 and fascism was making its entrance in Italy and Germany. This qualification is also important for those who, in retrospect, feel the compunction to re-write history. This also demonstrates Ortega's penchant and acumen for metaphysical essences and his overall good will and clarity of mind.

By offering such a pronounced warning, Ortega can go on to offer genuine answers to political problems at best or merely point out the importance of historical fact, at worst. Because of his philosophical concern for existential/metaphysical themes, he can write: "public life is not solely political, but equally, and even primarily, intellectual, moral, economic, and religious; it comprises all our collective habits, including our fashions both of dress and of amusement."[20] This is seen today as a rather stern warning, given that "post-modern" epistemology has displaced all semblances of the metaphysics of essence with an "all is political" approach to human reality.

19. *The Revolt of the Masses*, p. 11.

20. Ibid., p. 11.

The initial problem of the masses, Ortega goes on to say, is one of simple agglomeration. "This fact is quite simple to enunciate, though not so to analyze. I shall call it the fact of agglomeration, of 'plenitude.'" The initial stage of the problem is one of physical space, but more severely in importance, Ortega argues is that the masses now possess the instruments that technology furnishes them with without the slightest intellectual curiosity as to their origin. The sense of wonder that serves as the "luxury" of the reflective man and that creates the goods of civilization is quickly passed over in distaste by mass man, who is only concerned with practical value.

In all respects, mass man signifies a diametrical opposition to the virtues of the life of reflection. However, a clarification in this regard seems warranted: the life of reflection is not necessarily equated with the life of the philosopher. In fact, the main notion that Ortega explores in *The Revolt of the Masses* is that the masses by definition lack autonomy. Self-reflection is one of the safeguards of human autonomy. That the masses should pass this virtue off as too demanding does not take Ortega by surprise. What does seem new is that today the masses have become of one mind, where before this collective phenomenon existed in isolation. What has occurred, Ortega argues, is that what used to exist as man's metaphysical concerns have now been institutionalized as forms of politicization. The mass mind now understands itself to be the recipient of rights and privileges without any allegiance as to the notion of duty. Ortega explains: "Not only in any direction, but precisely in the best places, the relatively refined creation of human culture, previously reserved to lesser groups, in a word, to minorities."[21]

This is an early crucial point in the trajectory of *The Revolt of the Masses* where Ortega introduces the notion that "society is always a dynamic unity of two component factors: minorities and masses."[22] The "average man" is the commonplace, slipshod mind who signifies a degenerative moral and qualitative outlook on life. Ortega's clas-

21. Ibid., p. 13.

22. Ibid., p. 13.

sic text becomes confusing for today's reader who is accustomed to viewing all reality through the vague generalizations wrought by lazy social-political analysis. When Ortega speaks of mass man as lacking the ability to become "differentiated from other men," he does not squander this opportunity in yet a further attempt to "legitimize" human existence on the basis of vacuous political categories. This is precisely the problem of "revolt," as he sees it. Instead, the generic type called mass man now finds an impetus to carry out its revolt, given the protection and solace that it receives from politicization. Mass man's revolt, Ortega tells us, can be explained as a qualitative phenomenon because it naturally gravitates towards the greatest common denominator, where it can effectively release its tension and fuse it with that of the others. Yet what matters to the mass mind is not the like-minded others in the group, but rather the notion and perception of the group as refuge. Ortega writes:

> In those groups which one characterized by not being multitude and mass, the effective coincidence of its members is based on some desire, idea, or ideal, which of itself excludes the great number.[23]

The significance of this particular clarification is that Ortega designates mass man as a quantitative phenomenon — the existential/moral concerns of human existence as a "we" phenomenon. In opposition to this herd mentality, the coming-together of minorities takes place out of sheer coincidence. Philosophically, what Ortega attempts is a re-construction of the meaning of the self, especially in relation to others. The key element in this analysis is the introduction of Ortega's notion of conviction, where the latter stands alone and is willing to interpret reality at every instance. The significance of this statement lies in that "the select man is not the petulant person who thinks himself superior to the rest, but the man who demands more of himself than the rest, even though he may not fulfill in his person those higher exigencies.[24]

23. Ibid., p. 14.

24. Ibid., p. 15.

The Effort Toward Perfection

A clear exposition of Ortega's objective is witnessed when he writes: "The decisive matter is whether we attach our life to one or the other vehicle, to a maximum or a minimum of demands upon ourselves."[25] The *Revolt of the Masses* easily could have been titled "The Metaphysics of Strife and Resistance," because Ortega's analysis throughout this timeless work is no less than an exploration of universal essences. A fine example of this can be seen in his dissection of the division of society into masses and select minorities. While he refuses to accept this division as a mere social-political construct, he advances the argument that the inherent differences found in man are morally qualitative. A necessary condition of this analysis is the flow that ensues between mass man and select minorities and the places and institutions that either type engenders. What is significant in this respect is that while both types are naturally predisposed to create or dismiss corresponding life conditions that pertain to their respective temperaments, the greatest indication of a time of decadence is "the predominance, even in groups traditionally selected, of the mass and vulgar."[26] Ortega argues that this process is exacerbated by the erosion of institutions and modes of existence that were created by select minorities, but that now have been taken over by mass man.

Some critics of Ortega's work have called his thought "elitist" but such arguments are predicated on the pre-meditated negation of essences, to begin with. Such "arguments" dismiss Ortega's metaphysical analysis as "class constructs" etc., while simultaneously neglecting to realize that these attacks are ideologically conditioned. Besides being a form of self-serving ideological ranting, these attacks fail to consider the abundance of empirical evidence that Ortega presents. One of the factors that make *The Revolt of the Masses* a compelling metaphysical treatise has to do with the date its publi-

25. Ibid., p. 15.

26. Ibid., p. 16.

cation. While privy to the formation and dominance of totalitarian oppressive systems like Nazism and Communism, Ortega refuses to allow himself to be taken in by *reductio ad absurdum* arguments like the positivist sociologists, who unconvincingly attempted to link Fascism and the rise of Nazism to Socialism and Communist totalitarianism. But this presupposes the existence of the effect before the cause. What Ortega undertakes instead is to preempt such hollow sophisms by demonstrating the real-world conditions that certain human types bring about. If philosophy purports to offer knowledge of human reality, it must do so by looking at the causes that originate the conditions of all things human. Ortega's metaphysical analysis in *The Revolt of the Masses* is essentially one that, like Dilthey's *geistewissenschaften*, looks at the underlying spirit that informs human reality.

What can be misleading in reading Ortega's seminal text is that today most readers come to it armed with the biases of blind "tolerance" and pseudo-sensitive and dishonest social-political notions that negate reality — "notions born in the café," as Ortega refers to them.[27] This hardly seems the appropriate cultural climate to understand the thought of one whose sole approach to philosophy is grounded in humanism and common sense.

Ortega's notion that "nowadays, everybody is the mass alone" is a clear example of the force of what he refers to as mass.[28] This ability to be one's "sole mass" is directly opposed to a genuine Ortegan definition of individuality. Instead, the problem of human life can be effectively viewed as biographical drama precisely because at the crux of what it means to be human we find the essence of differentiation. Hence Ortega's analysis of the essence of differentiation — (*ensimismamiento*) or authenticity — is what he calls the understanding of reality proper, while its opposite, *alteración* (inauthenticity), is the "commonplace."[29] Human reality presents itself

27. Ibid., p. 18.

28. Ibid., p. 18.

29. Ibid., p. 19.

as a fragmented whole even though Kant's *ding an sich* (the thing in itself), Ortega suggests, never tells the entire story.

While reality might present only a finite human face, the conscious aspect of reality that makes sense to us, Ortega argues, always presents itself to the select man. The degree of truth that is uncovered is commensurate with the subject's level of engagement. Hence Ortega conceives human history as an aristocratic phenomenon.[30] However, like so many other aspects of this masterful work, the term "aristocracy" comes with a disclaimer: "social aristocracy has no resemblance whatsoever to that tiny group which claims for itself alone the name of society, which calls itself 'society'; people who live by inviting or not inviting one another."[31]

What seems so significant in Ortega's idea of nobility is that the latter demands a "least common denominator" attitude for itself. But isn't this tantamount to a notion of the "road less taken?" And if so, how can this be incorporated into the exigencies of the modern world? But nobility need not scare our sterile "post-modern" sensibility.

Ortega's idea of philosophical vocation is no less than a noble, even heroic enterprise. While reality is passive, it takes an act of reflection to organize it, he tells us. But this seemingly inane and simple task places great demands on the subject. In section two of his essay titled "Essay on Aesthetics by Way of a Prologue" Ortega argues that the philosophical act is laden with responsibility. He writes concerning reality:

> This is characterized by its passivity: In experience we find ourselves amongst contents — sensations, representations, concepts — that are "given" to us. Philosophy does not want to degenerate into illusionism, it must do well to limit it meditations to what experience offers.[32]

30. Ibid., p. 21.

31. Ibid., p. 22.

32. José Ortega y Gasset. *Obras Completas*. Volume VI. P. 268 [my translation].

The difficulty in ascertaining a "definite" answer as to the nature of human reality lies in its fleeting or teasing translucent quality. But experience, or what can be described as the interaction between human consciousness and barren, mechanical processes, does not bear the burden of elucidation or clarity. If we erase human consciousness from the aforementioned equation, what we are left with is a series of processes that while possessing a level of "reality" in their own right cannot be called "experience" proper. Yet the positivism that forms the background logos of "post-modern" reality places the primacy of reality on material processes. The material processes are referred to as: social/natural environment, political reality and bland sociological statistics. "On the other hand, if what experience showcases were complete in itself," he continues, "this would do away with the need for the work of philosophy."[33]

Ortega's ideas on metaphysics have direct bearing on *The Revolt of the Masses* because he argues that "revolt" is itself a metaphysical concept. He does not deny that this particular metaphysical angle of human reality has always existed. In fact, he asserts this to be the case. What has happened, however, in the development of "revolt" is both the conglomeration of mass man and the ascent to power by the mass minded. In other words, Ortega views this dominant "post-modern" form of positivism as responsible for unleashing man's primal, raw energy.

Where before the conditions of human existence demanded a level of reflection that was commensurate with our state of understanding — completing the unarticulated input of experience — today we are moved by the relative nature, or transferability of knowledge. This metaphysical depth and scope of *The Revolt of the Masses* surpasses all merely sociological renditions of man. Also important is the understanding that Ortega goes right to the core of metaphysical, universal essences whereas simple sociological analyses fail given the latter's concentration on historicism. Ortega mentions in some of his writing that man has no nature and that

33. *Obras Completas.* Volume VI. Madrid: Revista de Occidente, 1964, p. 268.

therefore everyone must seek to make himself, existentially speaking. But this does not preclude him from arguing that truth is objective and that the nature of individual perspective is to seek and filter this subjectively.

To unify human experience into one cohesive entity is the work of reflection. However, in Ortega's thought the notion of reflection merits a disclaimer. Reflection, or what he refers to as vital reason, is never equivalent to pure reason. Vital reason, as a form of reflection, encompasses a greater degree of human reality than pure reason. In fact, he argues, that pure reason serves man in the capacity of a technical "advisor" to the quantifiable aspect of human life. Vital reason, on the other hand, divides human existence into its vital components. Amongst these, we must include existential, cultural, emotive and beliefs aspects. Ortega stresses that to reflect is always to fathom human existence as being a conglomerate of diverse and also contradictory angles.

Ortega views human reality as demanding, as a great chain of resistance that makes life as we know it bearable. Resistance, then, does not give away its secrets. Human reality as resistance plays itself out to fruition as a kind of system of rite of passage that can ideally ennoble us. To understand these conditions and to have the will to embrace this challenge is an accurate definition of nobility. But what remains important in this process is to recognize the differences between vital reason as a tool for human salvation and pure reason as a much narrower end, a limited staple of man's intellectual pre-occupation. Vital reason is an all-encompassing tool that is not readily limited by its level of application whereas pure reason finds itself framed by considerations that may lie outside of its internal constitution. But this just serves as a roundabout way of saying that human existence is much more complicated than any axiom or theoretical principle can encompass.

Nobility is also best apprehended by vital reason in another more pressing manner. Ortega offers a fundamental understanding of duty and its relation to rights. He suggests a notion of rights that is ennobling and refreshing, especially compared to that word's current common usage. He argues that what the declaration of human rights was intended to achieve was to situate man in such a way as to foster his self reliance. The problem, however, as Ortega views it, is that with this new-found freedom comes a dire responsibility: self-autonomy. Self-autonomy is intrinsically connected with a vision of reality which offers resistance to the subject. What evades most commentaries on the idea of rights is that the word "rights" entails an existential dimension. Ortega explains:

> Now, the meaning of this proclamation of the rights of man was none other than to lift human souls from their interior servitude and to implant within them a certain consciousness of mastery and dignity.[34]

Ortega suggests that in many respects what the declaration of rights has done is to place man back in an ontological condition that begs for self-understanding in a cruder, more primal existential confrontation with the world. He goes on: "Was it not this that it was hoped to do, namely, that the average man should feel himself master, lord, and ruler of himself and of his life? Well, that is now accomplished. Why, then, these complaints of the liberals, the democrats, the progressive of thirty years ago? Or is it that, like children, they want something, but not the consequences of that something? You want the ordinary man to be master. Well, do not be surprised if he acts for himself, if he demands all forms of enjoyment, if he firmly asserts his will, if he refuses all kinds of service, if he ceases to be docile to anyone, if he considers his own person and his own leisure, if he is careful as to dress: these are some of the attributes permanently attached to the consciousness of mastership."[35]

The operative phrase in the previous quote is for the masses to assert "their will on society." The reason that this forceful volitional assertion seems significant is because Ortega does not believe that

34. *The Revolt of the Masses*, p. 23.

35. Ibid., p. 24.

the masses can — this, by definition — assert their will in a coherent manner. What are left then are blind material forces in the guise of what he refers to as "inertial thinking."

The central tenet of the problem of the masses has to do with what Ortega calls the "rise of the historic level." This can be defined as the condition wherein "the ordinary level of life today is that of the former minorities, [which] is a new fact in Europe, but in America the natural, the 'constitutional' fact."[36] The rise of the historic level essentially has to do with the existential conditions of individuals as these become self-aware. Because Ortega does not vie for the haphazardly accepted "post-modern" explanation of everything in the cosmos as a direct or indirect offshoot of politics, he enlightens this particular theme with hard- nosed metaphysical arguments. This is evidenced in the fact, he argues, that mass man's negation of an axiological hierarchy negates the notion of metaphysical existential differentiation, while accepting one that is social/political in nature. His point is that those who view human reality solely in terms of social/political association and who demand a forced, coerced egalitarianism consequently admit to existential differences in their demands. The proof of this is that social/political systems that recognize "marked" differences in this respect also manage to stunt any further human development once in power. This is, then, the central debilitating contradiction in all forms of totalitarian utopias.[37]

Another interesting reason why Ortega views the rise of the historic level as devastating is his notion of personal character and vocation. Again, we must make clear that while today conversation on social/political reality is often entertained by trite and insincere talk of "equality," Ortega's genius anticipated this politicizing of human reality by concentrating on individuals. Questions of human autonomy and what Julian Marías termed "individual spontaneity"

36. Ibid., p. 24.

37. These utopias are really dystopias to all outside of the blinding glare of ideology.

in his seminal work on Ortega, *José Ortega y Gasset: Circumstance and Vocation*, make up the great bulk of Ortega's work. There are many fine examples of the metaphysics of human differentiation, but few seem as potent as Ortega's thought on personal autonomy. He explains:

> The age of twenty-six — the figure must be taken with a certain amount of looseness — is the moment of most essential departure for the individual. Up to that time he lives in the group and of the group. Adolescence is cohesive. During it, man neither can be alone nor knows how to be. He is governed by what I have called "the instinct of coetaneity," and lives submerged within the herd of the young, in his "age class." But at that point in the cause of his life, the individual sets out toward his exclusive destiny, which is, at its root, solitary. Each one is going to fulfill after his own fashion the historical mission of his generation. For each generation is, after all, no more than this: a certain mission, certain precise things which must be done.[38]

There is no denying that Ortega understood the historic revolt of the masses as posing a serious and destructive challenge to human individuality and autonomy. This is so much the case that "revolt" seems to be a revolt against the subject itself. The freeing or "liberating" force of mass movements, he reiterates throughout his work, is precisely nothing less than a turning away from the interiority of the self. And as such, here lies the import of this alleged liberation. Without recourse to the balancing influences of temperance, self-understanding and an autonomous allegiance to duty, the self becomes nothing more than a vehicle for nihilistic vulgarity. For this very reason Ortega argues that embracing difficulty and the imperial resistance of reality, is the prerogative of genuine nobility. This leveling of reality to its greatest common denominator serves to deface both the autonomous nature of reality and the dignity of man. In this respect the dignity of the latter lies in the full understanding and embracing of the former.

38. Marias, Julian. *José Ortega y Gasset: Circumstances and Vocation*. Translated by Francis M. Lopez-Morillas. Norman: University of Oklahoma Press, 1970, p. 322

Still another devastating analysis that Ortega offers as a critique of modernity is his understanding of decadence. Decadence as a rule, he argues, is a decline from greater to lesser. Ortega navigates through the exigencies of this word very carefully. For instance, the maker of carriages finds the early days of the automobile to be a decadent time because it signals the departure of a particular form of life. And for the maker of "amber mouthpieces this is a decadent world, for nowadays hardly anyone smokes from amber mouthpieces."[39]

If Ortega were to interrupt his questioning at that point, this would only amount to a relativistic interpretation of decadence. Instead, he carefully sails through this word with the understanding that if it is to signify something greater than the mere attitudinal succession of generations, an ampler definition must be sought. He arrives at such a definition when he advances the problem into what is essentially an existential amnesia of the past. He defines decadence thus: "Hence for the time we meet with a period which makes *tabula rasa* of all classicism, which recognizes in nothing that is past any possible model or standard, and appearing as it does after so many centuries without any break in evolution, yet gives the impression of a commencement, a dawn, an initiation, an infancy."[40]

The major philosophical reason that Ortega attributes as the cause of this historic anamnesis is the incessant, perhaps even neurotic desire to be at the height of the time. What this does is to discriminate the achievement of previous ages — especially those that directly paved the way for the present. This contempt is a form of arrogance that Ortega attributes to the false valuation of the masses, given their rapid ascent to power while lacking the existential vigor and substance that such a responsibility requires. This is also another instance of "post-modern" "liberation." Yet again, the outward signs of mass mentality point to the internal principles that have helped to bring it about. The drama of the mass man is to

39. *The* Revolt of the Masses, p. 34.

40. Ibid., p. 36.

demand that this ascent should take place as a collective phenomenon. Ortega argues that human existence is precisely the opposite: a solitary trek that finds itself differentiated as a subject.

He writes, "the world has suddenly grown larger, and with it and in it, life itself. To start with, life has become, in actual fact, worldwide in character. I mean that the content of existence for the average man of today includes the whole planet, that each individual habitually lives the life of the whole world."[41] But the negative qualities of this series of quantifications come as a detriment to vital life. The numbing range of "liberating" possibilities now becomes a source of existential paralysis, for some.

41. Ibid., p. 38.

Chapter 2. Ortega's Notion of Mass Man and Noble Man

Ortega's thought pays great allegiance to man's being-in-the-world. While on the one hand he emphasizes this communication as a corrective to disproportionate attention and isolation of the self, this pendulum can equally sway too much in the direction of the world. This is the condition that occurs with his notion of the "increase of life."

But while it is ever growing larger, this psychological expanse has also created an ever-narrowing vista of human existence. This unexpected turn is ironic. The effect of the increase of life is one of leveling our circumstances, that is, "the world around us."[42] As a result, our individual circumstances come to be viewed as consisting of everything for everyone. If the "world" is now my circumstance, how can I be said to belong to anything, given that my circumstance is interchangeable with that of others?

Confused by popular abstractions, disassociated from reality and historically ignorant, Ortega would argue, is the call of the status quo. Naturally, he finds a decadent time to be joyously oblivious

42. *The Revolt of the Masses*, p. 41.

to itself. And what is the spirit of an epoch, if not what is easily conveyed by the guiding spirit of its members? Thus Ortega cuts through trite and cliché political baggage by suggesting "it is useless to talk of decadence without making clear what is undergoing decay."[43] But taken at its very core, what is decaying "consists in a lowering of vitality."[44]

Ortega's understanding of decadence points out a significant paradox. A decadent age is one that recognizes itself as lacking the capacity to embrace the past. However, our age, he insists, not only does not care to remember the past, but also feels superior to it. This self-bloating precludes any view of itself as decadent. Throughout this analysis, Ortega insists that decadence is measured as a qualitative vitality and not as its opposite: quantitative embracing of the height of the times. But if an age is collectively incapable of viewing itself as decadent, then how can we even make sense of the word? This is one sense where Ortega proves the importance of the antithesis of mass man: noble man. If an age is incapable of self-understanding and sincerity given its stronger desire for "liberation," isn't this also clearly what happens at the level of autonomous differentiation in vulgar versus vital man? Not only is this the case, Ortega argues, this is the atomic level of decay that defines any age. An age that lacks "clear and firmly held ideals" is an age that cannot or will not seek inspiration from a prior one.[45] He explains:

> But the truth is exactly the contrary; we live at a time when man believes himself fabulously capable of creation, but he does not know what to create. Lord of all things, he is not lord of himself. He feels lost amid his own abundance. With more means at its disposal, more knowledge, more technique than ever, it turns out that the world today goes the same as the worst of worlds that have been; it simply drifts.[46]

43. Ibid., p. 43.
44. Ibid., p. 43.
45. Ibid., p. 44.
46. Ibid., p. 44.

We can easily counter the slipshod arrogance of the aforementioned with the cosmic and existential insecurity of the noble man as: "every man who adopts a serious attitude before his own existence and makes himself fully responsible for it will feel a certain kind of insecurity which urges him to keep ever on the alert."[47]

It is interesting to note that even as early as 1930, Ortega already foresaw the momentum that our "post-modern" sense of entitlement was gathering. He has referred to this as the symptom of the spoiled child. This gratuitous and liberating sense of entitlement or what amounts to the "rights to this...or that," has many causes including the integral defiance of cosmic principles of reality. But always, his answers attempt to exorcise the metaphysical cause, not the political effects. Another clear and penetrating example of this — and *The Revolt of the Masses* is replete with these philosophical insights — is: "can we be surprised that the world today seems empty of purposes, anticipations, ideals? Nobody has concerned himself with supplying them. Such has been the desertion of the directing minorities, which is always found on the reverse side of the rebellion of the masses."[48]

The problem of the masses, then, suggests a genuine desire to be structured, either culturally, morally or spiritually by noble ends. How can it be, if by definition this would negate the reality of the masses? The question is centered on the genuine explosion of possibilities that exist today at the service of life, and the existential confusion and sheer heaviness that such possibilities entail.

Part of the insightful nature of *The Revolt of the Masses* is Ortega's ability to cite the specific origin of the "revolt" that he saw taking place in 1930 as really dating back to the 1850s. In an essay from 1921 titled "Karl Vorlander's History of Philosophy," Ortega elaborates on the importance of differentiation. There he cites the 1850s as a time consumed by ideology, thus making most philosophy then a pseudo-philosophy. He compares the pseudo-philosophers of the

47. Ibid., p. 45.

48. Ibid., p. 46.

first two decades of the twentieth century as following in the same ideological currents of the positivism of the 1850s. These ideological "prejudices and ignorance" are the predominant characteristics of Ortega's contemporaries, he tells us.[49]

The significance that essay has in light of *The Revolt of the Masses* is that in the latter Ortega offers a metaphysical basis for the idea of differentiation in human reality. Several points of contention emerge in the latter that are worthy of attention. One of these themes is that of specialization, which Ortega develops brilliantly in *The Revolt of the Masses*. Of particular interest to philosophy is the development of the idea that specialization in science, for instance, creates a form of "ignorance" that negates the original spirit of science in its overwhelming desire for understanding.[50] The other major theme that envelopes this essay is that of radical skepticism and the negation of truth. Both of these notions prove to be essential characteristics of respective forms of life that mass man embraces today.

But the question of radical skepticism — not just a doubting attitude that destroys certitude — is an essential component of the "liberation" attitude of "post-modernity." In the chapter of *The Revolt of the Masses* titled "A Statistical Fact," Ortega sets out to demonstrate that a central problem of modern life is its abundance — and, in some respects, the asphyxiating array of choices that we enjoy. We find ourselves in a world that we have not created, but "instead of imposing on us one trajectory, it imposes several, and consequently forces us to choose."[51]

This predominant existential theme of Ortega's thought has been most ignored by critics of his work. But even more importantly, this has resulted in a failure to tie in the freedom expressed in modernity with the failures of the masses to make use of it. What the mass mind does make use of is not primal freedom proper, but rather its

49. José Ortega y Gasset. *Obras Completas*. Volume VI. "Historia de la Filosofia de Karl Vorlander," p.299.

50. *The Revolt of the Masses*, p. 297.

51. Ibid., p. 48.

negation: moral relativism and existential inertia. Ortega makes this amply clear: "To live is to feel ourselves fatally obliged to exercise our liberty, to decide what we are going to be in this world. Not for a single moment is our activity of decision allowed to rest."[52]

Ortega packs the one hundred and seventy-nine pages of *The Revolt of the Masses* with a dazzling array of the metaphysics of human reality and where this leads us to in the social/political foundations of the "post-modern" world. The originality and breadth of this work is that where most works of existential philosophy remain cognizant of the importance of the differentiated self, Ortega leaps beyond this necessary step to demonstrate the individual spirit behind our most cherished, but flawed, modern assumptions and ideological biases. In other words, Ortega amply demonstrates the inherently logical connection between the answer to individual existential demands and the exigencies realized by the rule of institutions. In doing so, Ortega also manages to refute altogether the collective pretentious and dismissal of the subject that "post-modernity" holds so dear. As a side-glance, Ortega also manages to showcase the many collective myths set up by the anti-humanism of "post-modernity." He goes on to write: "It is, then, false to say that in life 'circumstances decide.' On the contrary, circumstances are the dilemma, constantly renewed, in presence of which we have to make our decision; what actually decides is our character."[53]

It is precisely this respect and allegiance for the autonomous subject that puts off Ortega's critics — most of whom, without studying his work very carefully, neither understand him nor care to accept the validity of his reasoned conclusions. This rank refusal to accept the authority of reality proper, then, becomes the *modus operandi* of "post-modern" thinkers. This is the only manner in which ideology can subject reality to the mere whim of "committed" collectivists. Yet need we point out that this conniving manner of prostituting oneself has long ago removed itself from the spirit of

52. Ibid., p. 48.

53. Ibid., p. 48.

philosophical vocation? Actually, it is not even the spirit of those who, knowing nothing else, know enough to respect the inner will of human reality.

Mass man implements a world, a political program — in short, a reality — that cannot help but represent the inner emptiness of its source and inspiration. This is the primary reason why *The Revolt of the Masses* does not concentrate — it hardly bothers with notions of power and state rule, etc., — on notions that are the catch phrases of ideology. Ortega makes the argument that the human environment comes about as a result of inherent differences and metaphysical differentiation amongst people so convincingly that it cannot help but to elicit the ire of ideologues. This much he makes clear:

> All this is equally valid for collective life. In it also there is, first, a horizon of possibilities, and then, a determination which chooses and decides on the effective form of collective existence. This determination has its origin in the character of society, or what comes to the same thing, of the type of men dominant in it. In our time it is the mass-man who dominates, it is he who decides.[54]

Ortega cites the problem of agglomeration as being one manner in which the masses have come to rule Europe and how "the new generations are getting ready to take over command of the world — as if the world were a paradise without trace of former footsteps."[55] The author points out that the rise of technical proficiency in the nineteenth century offers a broader stage where mass man can operate. He argues that the abundance of possibility will, in time, become an existential scarcity that gives rise to a protracted decadence, or what he calls in no uncertain terms "the vertical invasion of the barbarians."[56]

Gabriel Marcel, too, has a name for mass man, the barbarian who seeks to implement his particular form of vulgarity on tradition-

54. Ibid., p. 48.
55. Ibid., p. 51.
56. Ibid., p. 53.

ally constructive structures: peasant-proprietors.[57] Hence, Ortega's concern with the statistical reality of mass man does not attempt to establish the sociological, implications of this quantitative fact, but merely to allude to the tide of history in the absence of spirit. Undeniably, *The Revolt of the Masses* creates some pronounced difficulties for those whose vision of human existence is that of practical generalizations and collective efficiency. The general level of complexity that informs all existential reflection can only become fully manifest in light of the lives of others. One cannot easily deny that this particular type of reflection frightens the majority of people. Most people see little value in modes of thought that remove us from our well oiled and even smooth existence. This compounds our problem in trying to understand Ortega's work because it offers a large number of commentators and critics a level of difficulty that resists generalizations.

In *En Torno a Galileo*, Ortega plots out such an "interior" vision of human existence. We cannot help but come to terms with his existential themes.[58] Especially important for *The Revolt of the Masses* is Ortega's questioning whether it actually makes sense to refer to man as *Homo sapiens*? His contention has to do more with will than with intellect and more with the solitary trek of the individual than with societal institutions that attempt to minimize or ignore the centrality of existential frameworks.

Ortega's grasp of the nature of knowledge far from embodying utility, such as is the case in "post-modern" usage, is instead a reality unto itself. When we reflect, as in the original Greek idea of *theoria*, what is at stake is nothing other than my salvation. The objective understanding of what we are capable of has a direct correspondence to my degree of engagement with reality. This engagement has everything to do with my desire for transcendence or what amounts to my saving my circumstance. Appropriately enough the aforementioned work has been translated into English as *Man*

57. Marcel, Gabriel. *Man Against Mass Society*, p. 105.

58. José Ortega y Gasset. *En Torno a Galileo*, p. 9.

and Crisis. The crisis at hand is no other than man turning against himself, and not against the other as ideological pundits have obsessed over this question for the duration of the twentieth century. Ortega argues that circumstances are inert. Instead, we must act on our own and as a result, our decisions can never become transferable. How, then, do we arrive at the sociologically fashionable, and frankly tired and trivial, notion that social conventions are solely responsible for man's development? Why the pretentious notion that man is a helpless feather in the wind, blown about pointlessly and mercilessly? Ortega answers this question by addressing the politicization of all things human. Let us take, for instance, his insightful understanding in *En Torno a Galileo* of what he calls the myth of some people's lack of convictions. The "convictionless" man is a myth because all life, even that of the radical skeptic who lives in a destructive blaze of total doubt, already embodies a form of convictions.[59] Of course, this does not preclude the possibility that all convictions are equal.

The Revolt of the Masses signals a turn outward away from the self, away from the metaphysics of the interiority that is man and into the emptiness of external processes, much the same as the turning away from the apparent *cul de sac* of dealing with material elements that Socrates recognized. Ortega's view of mass man is that of a radical "emancipation" from the self — from the weight of reality proper. This solitary and existentially burdensome weight is removed in the promise of a lighter and freer existence. This emancipation views a collapse of objective values as liberation from institutions that dictate the course of vital life. Thus the turn occurs quickly once that the "new" institutions are erected that liberate us from the natural existential inquietude of the human condition. Is there a glaring contradiction present in this manner of liberation? Consider what he writes in *Man and People*:

> Belonging to and forming an essential part of the solitude that we are, are all the things and beings in the universe which

59. Ibid., p. 11.

> are there about us, forming our environment, articulating our circumstance; but which never fuse with the "each" that one is, which on the contrary are always the other, the absolutely other – a strange and always more or less disturbing, negative, and hostile and at best unconcurring element, which for that reason we are aware of as what is alien to and outside (*fuera*) of us, as foreign (*forastero*) — because it oppresses, and represses us; the world.[60]

Meditations on Quixote is the earliest articulation of Ortega's thought. The work has Ortega's earliest personal style to inform it. As a work of existential Personalism, this work has no rival. Fortunately, like all of his work, *Meditations on Quixote* is not a treatise, but an essay that reserves the right not to attempt a pretentious exhaustion of its subject matter. Finding himself in the midst of a thick forest, the author quickly comes to the conclusion that man's greatest concern is with the nature of reality proper. As a work of phenomenology, Ortega clearly captures the essence of this metaphysically vital entity that knows itself as such. The forest is an oasis from the surrounding world, but to take this as the principal meaning of this work is to miss its importance and intent. What is important in Ortega's meditation is the reality that manifests itself as a conglomeration of trees, each signifying a particular angle on the total make up of things. And much as these individual trees might obscure our total view of the forest, this by no means destroys the unity therein. Unlike "post-modern" renditions of reality, where there is no alleged center, Ortega's Perspectivism, if indeed it can be referred to as such, can only make sense precisely because there is a center. The center may elude or become veiled for some, but its ontological status is what allows for its transparency in the first place. Ortega's philosophical vocation and mission hence is not forceful as is often the case with pedagogy and always with ideology, for instance. His work is best described as suggestive of reason, and how this is manifested in and guides human existence. Con-

60. José Ortega y Gasset. *Man and People*. Translated by Willard R. Trask. New York: W.W. Norton & Company, 1963, p. 50.

sider his opening in *Historical Reason* where he addresses questions of directionless man:

> According to mythological belief, as you know, it was the duty of fauns — those followers of terrible Pan — to howl prophesies and predictions from the fearsome depths of forests as those capable of deciphering their essential meaning. [61]

Ortega distinguishes between mere biological descriptions of man and the biographical/existential interiority that defines the bare bones reality of differentiated man. While the former is a stable form of quantification that describes man as a member of a given species, the latter takes human reality as a subtext of the former where man finds the incongruous nature of the self as radical reality. A clarification of the two seems justified at this point:

> 1. That human life in the proper and original sense is each individual's life seen from itself, and hence that it is always mine — that it is personal.
>
> 2. That it consists in man's finding himself (without knowing how or why) obliged, on pain of succumbing, always to be doing something in a particular circumstance — which we shall call the circumstantial nature of life, or the fact that man's life is lived in view of circumstances.
>
> 3. That circumstances always offer us different possibilities for acting, hence for being. This obliges us, like it or not, to exercise our freedom. We are forced to be free. Life is a permanent crossroads and constant perplexity. At every instant we have to choose whether now or at some future time we shall be he who does this or he who does that. Each of us is incessantly choosing his "doing," hence his being.
>
> 4. Human existence cannot be transferred to another. No one can take my place in the task of deciding what I am to do, and this includes what I am to suffer, for I have to accept the suffering that comes to me from without. My life, then, is a constant and inescapable responsibility to myself. What I do — hence, what I think, feel, want — must make sense, and good sense, to me. [62]

61. José Ortega y Gasset. *Historical Reason.* Translated by Philip W. Silver. New York: W.W. Norton & Company, 1984, p. 14.

62. Ibid., p. 14.

As an existential program, human existence entails for Ortega the cognizance of agency: a subject. This subject is the receptor of all that "occurs to me" and that "I am conscious of." In one sense Ortega suggests that the things that happen to an unresponsive agent, that is, a non-subject — a tree, let us say — cannot properly speaking "happen" as they do in human existence.

To truly make sense of the differences between the lives and worldviews of mass man and noble man, it is first essential to understand what Ortega means by human existence. Because Ortega's thought does not place too much emphasis on positivistic/sociological conditions alleged by "post-modern" man, we immediately encounter a fundamental difference between the notions: human existence and life. Human life is explained as something that occurs to me — a vitality that runs through me — but that cannot be said to be me, properly speaking. On the other hand, human existence is a self-aware reality that is a witness to biological life and the existential possibilities that it engenders. These differences are essential to the dissecting of the aforementioned human types and they help in explaining our social/political reality.

Ortega's notion of solitude is essential to human existence; it is a major component of human existence as radical existence. But solitude ought not to be construed solely in the physical sense of being alone or physical isolation. Solitude signifies an existential stance on human reality. This solitude conveys a strong sense of responsibility for one's agency. Consider how Ortega ties this to epistemological questions. We have already said that his notion of Perspectivism exists as a circle that is contained within a broader objective epistemological concentric sphere. When he states that "two + two = four" only when I "retire alone for a moment and think about it," this is not to be confused with relativism. This point is a major source of confusion in his work.[63] Instead, he emphasizes that what is true is always the case, to a subject that is open and thus receptive to truth in its totality. When a child is first exposed

63. *Man and People*, p. 59.

to the reality of numbers, say, in kindergarten, what takes place is transcendence from a dormant state of ignorance to an active understanding of what there is to know. The child who does not attend school or who is not exposed, not so much to mathematics as to the possibility of mathematics, remains passive, in ignorance of such a realm. However, the greater confusion in the matter is that some critics of Ortega's thought have imagined this to mean a form of *esse est percipi*. Authorship in Ortega's estimation does not entail a destruction of reality, but precisely the opposite, the existential need to understand. He explains: "But for the present let us make firm our knowledge that the properly human in me is only what I think, want, feel, and perform with my body. I being the creative subject of all this, that is, of what happens to me as myself; hence, my thinking is human only if I think something on my own account, being aware of what it means."[64]

It is easy to see how the social/political, what he refers to as the practical world, remains a secondary arena where man can play out his essential or substantive drama. And furthermore, if what truly matters is a primary conscious reality called man as radical reality, it then follows that this secondary reality comes about not out of some ill-defined material process, but from the level of engagement that individuals maintain with themselves.[65] In addition, Ortega's notion of authenticity (*ensimismamiento*) encompasses an internal drama that is played outwardly in the physical world. Even our bodies remain two-dimensional beacons that allow for our maneuverability through spatial/temporal realm, but that in the end remain only that, external to ourselves and to "others." In terms of mass and noble man, this serves as a substratum of what is in essence an internal drama that extends even to our own astonishment. Thus: "not only am I outside of the other man, but my world is outside of his: we are, mutually, two 'outsides' (*fuera*), and hence

64. Ibid., p. 59.

65. Ibid., p. 62.

radically strangers (*forasteros*)."[66] "Radically" means no more than the original reality or source of all other subsequent realities that "I" represent to itself.

What the casual reader encounters in *The Revolt of Masses* at first may appear as a social/political treatise on the nature of European life in 1930. To read this book without further recourse to Ortega's other works is tantamount to taking effects as causes and thus severely limiting and damaging our capacity for genuine understanding. Yet this very question brings up a central dilemma in Ortega's notion of mass man: that to know is to want to know regardless of the meandering difficulties that this task may entail.

Life Presents Itself to the New Man as Exempt from Restrictions

The ironic emancipation of the self that Ortega mentions is the final step in mass man's historical "liberation." The irony in this is that historically, all forms of "liberation" have been those of external causes: master/slave, political tyranny and technological progress, for instance. Ortega's philosophical acumen is displayed in the boldness of his analysis. When he writes that history can indeed be foretold, he clarifies this statement by conceding that what we can prophesize of the future comes to us as general understanding. He rounds off this understanding by writing: "But that is all that we in truth understand of the past or of the present."[67] What is so illuminating about this understanding is his knowledge that people who deny the possibility of knowing the general direction that the future can take do so on the basis of the notion that since the world is essentially run by blind external forces, and since these forces cannot be said to be guided, then all such knowledge fails us. Implicit in this denial is the idea that man has no control over his life and thus nothing of substance can be gathered from paying attention to

66. Ibid., p. 75.

67. *The Revolt of the Masses*, p. 54.

man and our values. This is a negation of vitality, imagination and the necessary will to ensure control over material processes.

A most telling passage has Ortega arguing that perhaps the entire history of man witnesses its culmination in the triumph of zombie-like mass man. About this he writes: "The whole of history stands out as a gigantic laboratory in which all possible experiments have been made to obtain a formula of public life most favorable to the plant man."[68] The causes for this predicament, or indictment, as the case may be, are several, but the outstanding ones are: 1) the growth of liberal democracy, and the rise of technical knowledge both taking place in one century. Ortega's vision is historical in the sense that he views history as naturally gravitating toward a prescribed end. This might not be a deterministic or teleological end, necessarily, but it does suggest that every subsequent age reaps the causes of a prior age as effects. What Ortega does indeed suggest is that human history shows a marked propensity toward ever-easier forms of life. What this says about the nature of work, leisure and our ambitious propensity for a better material condition might not mean what some critics of his work have imagined. The question gets truly complicated when we realize that is one thing for noble man to create and reap the fruits of such labor, but another to realize that mass man enjoys, transforms and degenerates the fruits of this, an alien labor. And yet Ortega argues that a return to a pre-nineteenth century form of life would be suicidal. While offering a diagnosis of the problem and its causes, Ortega always manages to argue for a solution. Ortega's indictment is severe: "If that human type continues to be master in Europe, thirty years will suffice to send our continent back to barbarism. Legislative and industrial technique will disappear with the same facility with which so many trade secrets have often disappeared."[69]

Ortega's description of the new order that the new man has inherited embodies a full-scale understanding of how the fabric of

68. Ibid., p. 52.

69. Ibid., p. 53.

modern life since the nineteenth century is interwoven by moral and social problems. Of course, as I have maintained throughout this exposition, the central and original conceptualizing of the problem is a metaphysical one. At the outset of the problem is always the question: What is man? To this Ortega answers that man exists as two types: mass man, which he also alludes to as a vegetative existence — plant man, he calls it — and then there is also noble man. Some misguided commentators of Ortega's work, especially those who arrive on the scene armed with the myopia of ideology, have completely misrepresented his work, as they have equally done a severe disservice to other thinkers who have argued as sincerely as Ortega. Regrettably, Ortega's genuine philosophical concern has eluded these critics.

At the very core of the question lies the metaphysical reality that informs different lives. The lives of people before the nineteenth century were difficult, marred by a physical and economic narrowness of opportunity that forced them to adapt. Each extracted different possibilities out of his respective circumstances: "For the common man of all periods *life* had principally meant limitation, obligation, dependence; in a word, pressure."[70] Resistance, or what Ortega suggests several sentences later as "cosmic oppression," served the role of grounding human reality in limitation, a framework that went beyond the juridical and social sense.

While it might remain true that material progress clears the way for material enlightenment in the form of the Industrial Revolution, we must nonetheless tread carefully through such intellectual waters. One reason that we must temper our interpretation of Ortega's thought in this respect has to do with his notion of the level of appreciation that mass man has for all forms of life. In some respects, one can argue that Ortega's *The Revolt of the Masses* is a humanistic meditation of respect for cultural and scientific institutions. Ortega's thought retains a level of awe for the nature of things. In this sense, he is a full-fledged humanist because he places man at the

70. Ibid., p. 56.

center of the known universe as an agent who is endowed with primal existential freedom. The truth that we seek is that which is commensurate with our own level of existential inquietude.

Ortega's analysis is never a simple case of realizing the potential of a given age by utilizing the resources thereof. The given cultural and technological capabilities and prowess of any age, he has argued, is determined by man's level of engagement with his existential fortitude. There is hardly any magic in citing external causes. These are determined or, at worst, are processes that man moulds for his well being.

Mass man, Ortega argues throughout, has touched on the sensibility necessary to appreciate the rhyme and reason of things. Thus, he writes: "[I]n fact, the common man, finding himself in a world so excellent, technically and socially, believes that it has been produced by nature, and never thinks of the personal efforts of highly-endowed individuals which the creation of this new world presupposed."[71] As Ortega is keen on explaining the differences between noble and mass man, he is also forthright in citing the qualities and virtues that allow for these differences: of noble man we can say that such virtues include an autonomous will, loyalty, respect and most importantly, existential inquietude. To mass man, Ortega attributes the following two qualities: "The free expansion of his vital desires, and therefore, of his personality; and his radical ingratitude towards all that has made possible the ease of his existence."[72]

The Spirit of Self-Sacrifice

Perhaps the strongest virtue of the noble man is his capacity for self-sacrifice. This is also probably one of the least understood of Ortega's notion of nobility, and consequently of the mass mind. The negation of self-sacrifice occurs because man sees no limitations to his every whim. Feeling himself, unmoved and unrestrained by any

71. Ibid., p. 58.
72. Ibid., p. 58.

condition that lies outside of his sphere of influence, he withdraws from himself and abandons himself to external circumstances. This condition is precisely the opposite of nobility, of sacrifice. On the contrary, this condition makes man feel invincible, argues Ortega, precisely because of the ignorance of his capabilities. It is this vacuous, inflated ego that robs man of a genuine, autonomous subjectivity. The unwillingness to embrace and experience our limitations is a central trait of the mass mind. One of the effects of this ignorance of limitation is the misguided belief that the world revolves around our whims and desires. However, this might be a fine description of selfishness, as he says of the spoiled child, of avarice or vanity, even, but it certainly does not describe genuine subjectivity. This is merely its antithesis.

Ignorance of our limitations and incapacity to embrace difficulty is the hallmark of social upheavals and instability. Granted, what manifest as social/political effects has its cause in metaphysical/existential discrepancies. What Ortega saw as ominous signs of a future collapse of traditional valuation, e.g. "post-modernism," was already a constituent part of the make up of the "liberation" that the mass mind sought already in 1930s Europe. Part of the failure of critics who call Ortega's notion of noble man "elitist" is that what he classifies as the greatest virtue of this type of being is the willingness to view others as superior, morally or otherwise. While the mass man looks out onto the world as an unlimited range of possibilities, the noble man instead first looks inward in order to appropriate the nature of his self. In doing this, the latter achieves solidity in his sense of agency as a subject, while the former merely becomes transparent to itself.

The noble man is sincere enough to view others as superior to himself. This notion of the other is an inversion of most existential notions of the other. The affirmation that the other can exist as superior to me serves as a clarification of my own abilities and limitations. When I admire others, I can do so for two equally sound reasons: I admire a given quality because I too possess this quality and realize the difficulty of attaining it; or I lack such a trait, but recognize its worth. Thus, nobility of mind is also one of heart, in that one is essentially prepared to sacrifice one's self and one's view of one's self in favor of a greater truth. But what happens in an age that negates the existence of truth and when state and institutional apparatus have been implemented that promise the eradicate strife? Ortega essentially argues for a form of catharsis that keeps the autonomous subject grounded within himself.

What exactly has changed in the world to warrant the birth of such an over-abundance of mass man? Ortega's essayistic style always reiterates, re-traces its steps, in an attempt to explain fundamental points that might not have been completely understood the first time around. Hence he asks this question several times throughout the work. What has occurred is that "the new world appears as a sphere of practically limitless possibilities, safe, and independent of anyone."

Ortega's philosophical acumen penetrates into spaces that today have been plugged up by social/political theory, but that were previously the domain of man's essential differentiation. He argues that future generations will be formed in circumstances of vacuous notions of reality the same way that the consciousness of prior generations was formed with an eye toward metaphysical resistance. What is striking here is the degree of reality that his thought process is able to unearth by simply remaining open to evident truths. He continues: "For that basic impression becomes an interior voice which ceaselessly utters certain words in the depths of each individual, and tenaciously suggests to him a definition of life which is, at the same time, a moral imperative."[73] The cue that people take from reality is not limited to the spatial/temporal spaces that they inhabit, but to internal essences. Reality cannot easily be denied, much less refuted.

73. Ibid., p. 61.

Another essential point to consider about *The Revolt of the Masses* is that it was written by what we call today a public intellectual. Ortega had no ties to universities or other institutions that might conceivably possess a dominant political agenda, either in his time or ours. It is also important that Ortega's work is not contaminated by the blindness that ideology wreaks on ideologues. Few "intellectuals" today are sincerely willing to embrace Ortega's sophisticated and insightful argumentation.

His mass man is akin to a child who has been let loose in a toy store or park. Where to begin? Ravaged by the "liberating" enjoyment of external forces, he loses himself in conditions that are not of his design but which he feels are merely at his disposition. The compulsion that this child/mass man feels must be continued from within by some interior necessity.

When noble man sacrifices himself he does so from "some standard beyond himself, superior to himself, whose service he freely accepts."[74] Noble man makes demands on himself that in essence force him to transcend himself. What seems so important in all of this is the acceptance that this attitude embraces all of the acts of this character type, thus a form of life. Hence, the life of noble man displays a self-transcendence that is evident in its ability for self-sacrifice. Far from demanding anything from others, the noble life instead makes great demands on itself. And thus Ortega defines nobility by the demands that such a life makes of itself, by the obligations that it must meet. Self-sacrifice embodies a respect for truth and moral obligation that is attuned and that is regulated by internal principles that such a life is compelled to ascertain. Self-sacrifice is the staple characteristic of noble man because it loses its effect on the understanding that this virtue is good in itself without seeking external rewards.

Ortega makes a sound distinction between private privileges and common rights (positive, legal). The former are metaphysical/existential in make up because they come about as the result of the

74. Ibid., p. 63.

necessary vision to lead a life of resistance to cosmic difficulties. What such a vision uncovers is the essences that, like a sieve, are not designed to contain the whole of human reality. The essences that give birth to private privileges are enacted without the consensus of the public view and by their nature eschew all institutional review. Legal rights, on the contrary, Ortega tells us, are passive entities for mere enjoyment that "do not answer" to any particular effort on our behalf. Common rights, "the rights of man and citizen" are static entities that, without a sincere level of engagement, appreciation, and reflection, become ossified. However, Ortega's contention is not to minimize the importance of the aforementioned rights, these are attained by the effort of others in a prior date, but that the person's upon who these rights are bestowed need to actively embrace them as originating from their effort.

Even though nobility is not a fixed, constant trait of man's, Ortega views this as the fundamental form of life of noble man. But maintaining this required level of existential awareness of one's life and the world around us is a life-long process. And yet the life of reflection is not a life of over-analysis or hair-splitting pedantry, either. A permanent fixture of this form of life is that it is not as natural to man as some commentators, Aristotle included, have suggested. Ortega views man's rational capacity as being a tool that man can use when needed. But this is not an end in itself. The noble man is besieged by this constant need to ascertain the meaning of personal experience. This only adds to the list of major differences between this character type and mass man.

Not much has been said by commentators of Ortega's work about the correlation between Ortega's notion of noble man and a stoic attitude toward life. He argues that nobility does not seem to make its mark in ancient civilizations until the Roman Empire, and even here it is used as a negation of hereditary nobles. The stoic ideal in Ortega's work serves as an antidote to the greatest common denominator in man: biology. The contrast between the two is startling. From the outset, it is clear that in Ortega's thought this stoicism removes man from a mere biologism.

But given Ortega's respect and desire to invigorate human existence with a broader understanding of life, perhaps a more appropriate description of mere biology is that of an aberrant sensuality. Without a doubt, a major component of Ortega's critique of "post-modernity" is the latter's unleashing of an all-embracing sensualism. This sensualism is passive, embracing the latest fashion and trying to outdo itself in attaining to an ever-greater level of the times. This also makes this sensualism into a reactive mode of life that envisions that the grass is always greener somewhere else without convictions or a prescriptive vision for life. This sensualism that can only exist as the corollary of an invasive keeping up with the level of the times is said to be centralized and thus dictates the course of the cultural, technological, moral and social/political. For this reason, it is important to separate mere biological life from the existential category: human existence. Whereas the biological is akin to the sensual, human existence recognizes the ontological foundation of what it means to be — and then — what it means to be oneself.

Chapter 3. Subjectivity and Mass Culture

The main reason that *The Revolt of the Masses* surprises some people and confuses others is that while the title contains the world "revolt" — raising expectations of a social/political kind — the book is truly a work on the metaphysics of human life. More specifically, the book is a meditation on questions of an existential nature. More than seven decades after its publication, *The Revolt of the Masses* proves its major contention: that "post-modern" man cannot conceive of life in other than social/political terms, given the contradictions that this "theoretical" liberation creates. The moral/intellectual poverty of our present-day predicament does not know what to do with such a text. *The Revolt of the Masses* is not a formulaic work, and as such it does not espouses the fashionable and calculating, social/political invective of ideology. Much more, this is not a predictable work. Nothing contained in its pages speaks in the conditioned, political correctness, disingenuous doublespeak that is practiced today. Ortega means preciously what he says.

This is the kind of work that forces commentators, critics and "theorists" to come up to the level of its author. *The Revolt of the Masses* is not a pretentious work that is artificially replete with sociological "data." Ortega's respect for reality finds us today much like young children who must be reprimanded for public indiscretions. We find many clues that prove this throughout the work. Unfortunately, a large portion of his thought has still to find a home in translation. However, much of his work that has been translated into English bears out the major thesis of *The Revolt of the Masses.*

One such example is *The Dehumanization of Art and Other Essays on Art, Culture, and Literature.* In his insightful essay, "In Search of Goethe from Within: Letter to a German," Ortega develops some of his seminal existential themes: life as effort and man's inner destiny. This is a work that dates from 1925. This fact is in itself important for at least two reasons: it predates a very significant portion of the work of other existentialists and it also serves as an example of the cultural, literary or philosophical themes that Ortega explores. It also serves as a fine example of the themes that Ortega views as replete with existential importance. This same idea makes up a large theme of The *Revolt of the Masses*, where what matters is the inner life of man, not ideology.

In Search of Goethe from Within

When Ortega argues that destiny is not chosen he is addressing not the external circumstances of man, but rather his internal constitution. The essay on Goethe is indicative of an attempt to understand the Other from within and not just some overt form of criticism or evasive psychoanalysis. Ortega's originality and clarity of intent in this respect is an attempt to view the Other as an existential subject. This concern of his has been alluded to as life as narrative, life as biographical, and life as vital reason. Without embarking on a discussion of these themes at the present moment we must, however, concentrate on his notion of the interior reality that life is for itself.

Ortega's essential idea here is to undertake a study of the interior life of the Other, not as an outsider would imagine him but as a genuine and sincere attempt to witness what another life might be like. This, of course, presupposes several things including the understanding that, for this process to succeed, a commonality between these two subjects must already exist. This means that one must be up to par with the Other if we are to witness their life as they do themselves. This has immense repercussions for amongst other things: Ortega's essay's on love, literary and artistic criticism, and his ideas on reading. One of the strongest aspects of Ortega's thought is its degree of genuine sincerity and clarity. He explains: "Man recognizes his I, his unique vocation, only through the liking or aversion aroused in him by each separate situation. Unhappiness, like the needle of a registering apparatus, tells him when his actual life realizes his vital program, his entelechy, and when it departs from it."[75] This impetus to realize oneself is what Ortega calls "tragedy." To recognize the above mentioned "vital program" is already an indication of a proactive mode of existence. Finding himself in the broad expanse of the material world, man is then forced to take shelter in himself except that this cure comes at an immense price. He continues:

> Man, with all his preoccupations and efforts, is delivered over to the outward, to the world around him to the extent required by his ends. But concerning himself he knows only when he is satisfied and when he suffers, and only his sufferings and his satisfactions instruct him concerning himself, teach him what to seek and what to avoid. For the rest, man is a confused creature; he knows not whence he comes or whither he goes, he knows little of the world, and above all, he knows little of himself.[76]

The interaction between what Ortega calls our "life-design" and our actual life is the sum of the strain of human existence. This is the source of our ailments as well as our happiness. Ortega makes this very clear when he argues that for plants, animals, or stars this strain does not exist. These forms of existence are fixed and locked

75. José Ortega y Gasset. *The Dehumanization of Art*. Translated by Helene Weyl. Princeton: Princeton University Press, 1968, p. 152.

76. Ibid., p. 152.

in a determined order. Hence in the life of man tragedy is explained as man's voluntary succumbing to an inevitable fate. But tragedy as such exists only in the instances when the inner destiny of individuals cannot be realized. The failure to fully articulate our inner destiny given the objectifying forces at work in the universe is what Ortega designates as the drama of human life. This aspect of life is never dramatic in a theatrical sense. Instead, it is more akin to the likes of a monologue — the rational/existential dialogue that takes place within the self.

The chapter on Goethe in *The Dehumanization of Art and Other Essays* is an instructive example of the range of existential themes present in Ortega's work. Goethe, along with the superficial grasp that his biographers had of his inner world, according to Ortega, serves as a solid ground from where to demonstrate his notion of the resistance and strain that the external world exerts on our lives. But this is only one side of the equation. The other has to do with this same resistance and strain that is allowed a prominent place in our lives through our own free will. This, Ortega suggests, is the epitome of "to fight oneself." Sartre called this condition bad faith.

Ortega's idea of biography is consistent with the revolt of the masses because his greatest contribution to this question turns out to be an existential analysis of man revolting against himself. One of the components of this revolt is the obvious damage that people do to themselves by turning against all semblances of an inner vocation. Of course, nowhere does Ortega assert that everyone has a vocation. Self-revolt only applies to those who do and refuse to act on this self-understanding. Goethe, Ortega argues, was perhaps a fine example of this biographical evasion.

Vocation is always individual, but as such individuality can only be what it is because it exists in a multitude of things. Differentiated from other external entities, vocation seeks to rule over a given life. But this is merely another way to suggest, "When we speak of life, every word must be completed by the appropriate index of individuation. This deplorable necessity is indeed a part of man's destiny as man: to live "in particular" he has to speak "in general."[77]

What often makes Ortega's idea of vocation such a difficult notion to grasp is that he does not view vocation to be commensurate with our most indisputable gifts. While our personal gifts allow us to maneuver through particular aspects of the world, they only serve as the truncated examples of the axis where vocation originates: Vocation can be explained as the culmination and exaltation of a given life. Part of the reason that vocation is ignored or displaced as the central component of inner life has to do with an existential lack of security. Ortega argues that to try to comprehend the essence of human life is also an attempt at comprehending the existential insecurity therein, thus man comes to the realization that the promised security of the external world often comes as a falsification of our vocation. This means that whatever comfort and firm ground our understanding or vocation may bring to our lives, it nevertheless forces us to recognize that the next minute becomes as insecure as the last. Naturally, where mass man is concerned, Ortega's existential characterization of human life cannot help but scare away the many. How many people want to be reminded that life is often an existential burden? In the event that people come to understand this difficult truth, we must still ask ourselves just what will most people do with such unbearable understanding? Ortega explains: "When we speak of life, every word must be completed by the appropriate index of individuation. This deplorable necessity is indeed a part of man's destiny as man: to live in particular he has to speak in general."[78]

Ortega's answer involves the recognition that such difficult and fundamental truths only play a negative role, that is, their effect is felt as negations in human life. All misguided forms of "liberation" act as if the former is true and thus there is nothing that can be done about it anyhow. Why struggle against these cosmic principles, as

77. Ibid., p. 152.

78. Ibid., p. 159.

Ortega refers to such truths, if I am defeated before I begin? This, then, is the response of the masses, Ortega tells us: a flight from the difficult, the sacred and what characterizes us as existential beings in the first place.

Life as Reflective Task

Ortega offers a philosophical treatment to a topic that writers, painters, sculptors and other artists have made the source of their inspiration: the question of the meaning of life. The forms of personal expression that come as a result of people's motivation to live and act is the stuff of art and even science. But such personal expressions, when they embody a reflective stance, are equally what promote well being in the life of farmers, gardeners, explorers and just about anyone who feels the need to reflect. But here, too, it becomes difficult to take seriously the question of meaning, given our "post-modern" trivializing of the sacrosanct. When today's aberrant and lowly popular culture takes up questions like the meaning of life it does so in such a trivial and insipid manner that no genuine answer is sought, and the question itself is no longer taken seriously.

"Post-modern" nihilism has converted all seriousness into rhetorical questions. An appropriate question to ask at this point, then, is: what will happen to a mankind that cannot fathom serious inquiry concerning life and death? This is significant because we cannot pretend that suffering, mortality, despair, joy and the existential uncertain quality of human existence will be "conquered." What this spirit of nihilistic liberation will accomplish in the short run is undoubtedly a truncated manner of "settling" these concerns by obscuring them altogether. Ortega is correct in arguing that the concerns of future mass man will be those that carry political, thus public, rewards, or what Camus referred to as "all of the modern mouthwashes."[79] He seems to ask: Why expose ourselves to private suffering when we can transform this into highly visible and

79. Philip Thody. *Albert Camus: A Study of his Work*. New York: Grove Press, Inc., 1959, p.363.

rewarding public causes? When sinking in the bog of existential uncertainty, take comfort in the higher ground of the politically expedient and officially sanctioned.

The Revolt of the Masses comes complete with the elucidation of profound ironies. While in some quarters man is reprimanded for not possessing "ideas," *The Revolt of the Masses* argues that an overabundance of "ideas" is precisely the problem and not the solution. He is careful not to confuse idle talk with genuine thought. The real concern, as Ortega sees it, is that while existential vitality has dwindled, so too have ideas turned into word games. This last point is easily verifiable today by the definition of "post-modernism."

When Ortega argues that we "must stop living from our ideas and learn to live from our inexorable, irrevocable destiny," he means that what truly drives a centered life of convictions is self-understanding and not public ideas.[80] He completes this thought by stating, "Our destiny must determine our ideas and not the other way around."

Ortega's logic, in terms of the convictions of mass man, is easy to follow: if an entire culture lives and utilizes the thought of others (noble man) then such thought should be recognized as the vital embodiments of their originators. What mass man does with these thoughts is evident given the framework of Ortega's book: the masses vulgarize culture. He explains:

> Primitive man was lost in the world of things, there in the forest; we are lost in a world of ideas that shows us existence as a cupboard full of equivalent possibilities, of things comparatively indifferent, of *ziemlichgleichgultigkeiten*.[81]

Ortega's critique of ideas has to do with the objectification and politicization of culture over life. The problem, as he sees it, is that we have become too conscious of consciousness itself. As such this is a false, abstract or hyper-consciousness that does not know limitations. It is for this very reason that Ortega views revolt as spring-

80. *The Dehumanization of Art*, p. 169.

81. Ibid., p. 169.

ing from a moral immaturity based on a false assumption of existential security, the dominant malady here being a superficial notion that "all is possible." The obvious outlet for this vacuous feeling of incivility is that of an all-embracing, superficial and destructive youth culture. What can be more in keeping with "post-modern" nihilism than institutionalized indiscretion?

The natural course of events in the life of the reflective youth, Ortega argues, has always been one where "the growing insecurity of his existence proceeds to eliminate possibilities, matures him."[82] In the absence of this insecurity he will never cease to be a youth even when this goes against his circumstances. Ortega sees this as a flight from the self, where eventually "the struggle with the rest of mankind begins," and where "asperity, bitterness, the hostility of our mundane environment appear."[83] This is a crucial moment in the formation of the mass mind — a total revolt against cosmic logos — and the embracing of a heroic stance.

The Heroic Stance

Animal life is pure *alteración* (inauthenticity), an entity lacking an internal dimension. When an animal is in flight from some danger, it flees to another location that it perceives as a safe haven, at least temporarily. The best that animals can do, then, when their life is in danger, is to seek the relative comfort of another locale. Thus, it becomes easy to conceive of animal life as perpetual motion, uneasy at that.

In an essay titled "The Self and the Other," Ortega cites Baudelaire's answer to the question, where he would like to live? to which the French writer answers: "anywhere, so it were out of the world!"[84] Ortega's point here is not social/political but a clear example of what he means by the interiority that is representative of the self.

82. Ibid., p. 169.

83. Ibid., p. 170.

84. Ibid., p. 180.

The Spanish word *alteración* means outside of oneself, an inner state of "tumult" or the loss of autonomous control. Ortega's use of this word is tantamount to an existential category. The importance of this condition to his analysis of mass man is paramount to a flight from the self into the world of the other and the external conditions that create a secure web of noise — Ortega calls this tumult — around the frightening inner silence and solitude of interiority.

Like the lower animals, when man flees a perceived danger he too flees into the comfort of a protective environment. An obvious difference, however, between man and the lower animals is that man does not encounter danger based merely on his lack of protective camouflage, or his inability to fly, swim or run in any given situation. Unlike the animal, man creates a world around itself — his circumstances — that can be of his own doing, such as the anonymity sought in societal conventions. In *alteración* — and today we have some very innovative and modish ways to suggest this, including the denial that this phenomenon ever takes place — what we flee from is our own internal constitution. Finding the droning silence of differentiated autonomy too much of a cross to bear, we flee from ourselves. But where can we go once that we can longer fit within our skeletal space?

What characterizes the noble life, Ortega reiterates throughout this work, is the ability to abandon the external world and enter into oneself. Yet this is misleading because while Ortega conceives reflection and introspection as uniquely human abilities, he does not suggest that these are easy tasks to enact. They merely exist as potential, possibilities that require a great deal of energy and strife: "nothing that is substantive has been conferred upon man. He has to do it all for himself."[85] The ability for introspection is a privilege, he tells us. But this is a privilege that organizes the order of human existence around a three-dimensional being.

The ability to remove himself from the affairs of the external world for extended periods of time enables man to reflect on his

85. Ibid., p. 183.

circumstances and refresh his vital energy. But most importantly, this retreat into the self grants man the power to form ideas.

These newly formed ideas allow for a "plan of attack against his circumstances."[86] Once man "returns" to the world, he does so with convictions that have been mulled over with care. By all indications, according to Ortega, this is an activity that is rejected by the mass man. But because man is of the world and has been firmly rooted in its exigencies, the ideas that man forms can never be those that respond to mere personal whim.

Instead, for ideas to contain any degree of cogency they must bear what Ortega views as a "reference to the behavior of things." His point is that if our ideas are going to be formed by a denial of reality, then it would seem futile to argue that man must coexist with the external world. The usefulness of ideas resides in that they say something about reality.

The reason that mass man refuses to concentrate his energy in the task of self-reflection is because this task is as alien to man as speech is to the weeping willow. He explains: "This inwardly directed attention, this stand within the self, is the most anti-natural and ultra-biographical of phenomena."[87] Ortega suggests that man's lack of power to concentrate and his ability to become easily absorbed by external impressions makes man an eternal quasi primitive who must strive for whatever self-understanding that he can settle for. "Primigenial" man is the name that Ortega uses to describe this type of mass man who is perpetually submerged in the things of the world.

Ortega equates a turn inward toward the interiority that is the self as the supreme act of heroism. What makes this act heroic is demonstrated in the sheer resistance that this task involves. Another aspect of this supreme form of heroism has to do with its inherent ability for action. When man retires to reside within himself (*ensimismamiento*), he only does so in order to plan or orient his fu-

86. Ibid., p. 184.

87. Ibid., p. 186.

ture action. Thus action cannot be conceived as such without the foundation of a prior reflection. Action without thought is only raw energy.

Having said this, we must make clear that man, according to Ortega, does not live to think, instead thinks in order to survive. Thought is not a luxury in this sense but a necessity that lies at the core of human salvation. The essence of heroism lies in our embracing a life of difficulty. Thought is drama because it is dynamic, ever on the verge of being vanquished, lost.[88] At the personal level — and this is only the crux of importance in Ortega's work — man needs to remake his existence daily in light of his existential insecurity.

At the cultural level, however, Ortega argues that man has become degenerate because he has become accustomed to the belief that progress is both necessary and inevitable. This has led man to believe for the first time in history, Ortega contends, that life is secure: "The fate of culture, the destiny of man depends upon our maintaining that dramatic consciousness ever alive in our inmost being, and upon our feeling, like a murmuring counterpoint in our entrails, that we are only sure of insecurity."[89]

The repercussions of this pseudo existential security for man were particularly felt in the last century, with the material progress that has eased man's condition. But here Ortega is not denouncing progress necessarily. He views the latter as inevitable. What he is concerned about is just how this modern condition has come about.

88. Ibid., p.186. "The intellectualist aberration which isolates contemplation from action was followed by the opposite aberration — the voluntarist aberration, which rejects contemplation and deifies pure action. This is the other way of wrongly interpreting the forgoing thesis, that man is primarily and fundamentally *action*. Undoubtedly every idea, even the truest, is susceptible of misinterpretation; undoubtedly every idea is dangerous; this we are obliged to admit once and for all, but upon condition that we add that this danger, this latent risk, is not limited to ideas but is connected with everything, with absolutely everything, that man does. Hence I have said that the substance of man is purely and simply danger. Man always travels along precipices, and, whether he will or no, his truest obligation is to keep his balance," p. 189.

89. Ibid., p. 191.

In other words, Ortega is pointing out how we have arrived at the point where we find ourselves today.

Vacations from the Human Condition

If no genuine action can be achieved without a prior plan that is born in contemplation, as Ortega asserts, it then becomes clear that "post-modernity" is no less than the freeing up of man's raw, ill-guided animalistic nature. Genuine action is born invariably in its having been considered first. However, this does not entail that thought is the highest value in its own right. But reflection is a form of solitude. Perhaps even the ultimate form. And it is in solitude that man can reconstruct the meaning of his individuality. The noise of the "post-modern" world can be measured by the advanced state of primitivism of what used to be musical form, for instance. The hyper visual condition of today's cinema, the endless array of public opinion and "discourse" where everyone must be right — in theory: this condition serves as nothing less than a colossal arena for *alteración*. Where in all of this attenuated sensualism does one encounter silence and solitude? "Post-modernity," in Ortegan terms, can be defined as the creation of a world where man is "liberated," to such an extent that he turns against himself by embracing a destructive nihilism. This type of man finds himself beside himself and never comes to suspect this condition. Ortega refers to our age as an age of blind action that produces a concatenation of stupidities and cultural inanities. Ours is an age of "action" where all forms of life have been politicized. This is a condition where the venerable Eric Hoffer argues, in *The True Believer*, that people becomes ripe for all kinds of mass movements:

> There are other safer substitutes for a mass movement. In general, any arrangement which either discourages atomistic individualism or facilitates self-forgetting or offers chances for action and new beginnings tends to counteract the rise and spread of mass movements.[90]

90. Eric Hoffer. *The True Believer: Thoughts on the Nature of Mass Movements*. New York: Harper Perennial Modern Classics, 2002.

Today's new form of life — and the explosion of the "liberated" new man on the world scene — has submerged the life of the individual under a diverse array of artificial social/political forms of life that has destroyed the closeness that man previously felt to itself. The new man is content to shoot destructive salvos at himself while calling this liberation. This is what Ortega calls an inability for happiness, precisely because "man is condemned to an inability to be substantially happy if he cannot be happy in the style of his own time."[91]

Silence, solitude and a life of reflection are all themes that Ortega develops throughout his writing, beginning with an early essay titled "Adam in Paradise" and following through quickly thereafter with his first book, *Meditation on Quixote* and later *Meditations on Hunting,* to cite only a few works. These themes, like everything else in Ortega's work, are never taken in isolation. For example, some of these themes elicit the meditative silence of *Meditations on Quixote* and the reflective stance that man takes toward society in *Man and Crises.* On other occasions Ortega does not take these themes head on, but instead takes a suggestive, nuanced view of their importance. This latter treatment of philosophical themes eschews pedantry, preferring essayistic clarity and readability.

In *Meditations on Hunting*, a book that takes as its theme the ancient art of the hunt, and as its form a meditation on human existence, Ortega concentrates on themes like life and death. Ortega's careful handling of these themes becomes the *de facto* manner of writing philosophy. Absent are the neologisms, specialized technical jargon, and the self-referential hair splitting that alienate conscientious readers. The number of educated people who read Ortega and who do not possess specialized knowledge of the subject is truly astounding and a great tribute to the power of philosophy as a discipline. Since his death in 1955 Ortega has come to be recognized as a profound cultural voice throughout the world, but especially in Spanish speaking nations.

91. *Meditations on Hunting*. p. 114.

Meditations on Hunting uses the ancient art of hunting and conservation to locate man's place in the cosmos. Hunting serves as a symbol of man's coming face to face with his nature both, as an archaic past and as a possible future. Even though Ortega argues that man cannot return to a previous age, he does state that the pathos of any given distant period can be embraced biographically in order to make sense of the present. By returning to the values of hunting, Ortega argues, that man can suspend his dependence on the outward, material circumstances that he has come to take for granted. The greatest threat that mass man faces is his inability to build his life around individual autonomy. The irony in this condition is that a retreat or diversion away from the world is necessary for mass man while an immersion in the world is a luxury that noble man indulges in as a result of a life of reflection. Here we think of the poet's assertion that the world is too much with us.

But just what can it mean to take a vacation from the human condition? Ortega's contention is not that taking a vocation from the world is the main problem of the masses. Actually, this is the understanding of leisure that the noble man is privy to. Noble man takes a vocation from his external circumstances because this allows him to come to know himself, to refurbish his drained spirit. The "vacation" does accentuate his ability to come to terms with his existential makeup and vocation. Keeping away from the background noise that is the "post-modern" world garners the reflective being the quality of an inner life. Even the trials and difficulties encountered by the primitive hunter, "crouching," "concealment" or "disguising" oneself in animal skins, Ortega points out, is a form of emitting the animal, of learning from the external circumstance. But this can only be assimilated into the human condition by careful study of the animal, by paying allegiance to detail, both, qualities that can only be achieved but through a reflective process.

Given the absence of this reflective stance, mass man must receive its well being from factors that are not of its own doing. However, one of the greatest stumbling blocks of mass man is his refusal to acknowledge any greater authority than himself. This " I won't allow others to accomplish what I cannot achieve myself" attitude is a central component of mass man that requires a section unto itself to which I will return.

The retreat from the world that is so characteristic a trait of noble man serves as a safeguard for values that otherwise lose their luster in lieu of keeping abreast of the "level of the times." Ortega fully recognizes that the meaning of keeping up with the times is the greatest force that "post-modernity" exercises in alienating people from themselves. The reflective, and we must add, vital act of the noble man's enterprise is not unlike Descartes methodological doubt. Most people who have attempted such a method as a way of life understand the difficulty of this task. The greater part of the difficulty in this task is that Ortega, as well as Descartes, seeks certainty, a task that requires sincerity and intellectual integrity. This is the purpose of their doubting. Neither thinker is a skeptic. Also, in many respects, Ortega is asking to what degree the "post-modern" world is the exact opposite of the stoic life that embraces personal strife? This thought is brilliantly conveyed by C.S. Lewis when he writes in *The Abolition of Man*: "What is now common to all men is a mere abstract universal, an H.C.F., and man's conquest of himself means simply the rule of the conditioners over the conditioned human material, the world of post-humanity which some knowingly and some unknowingly, nearly all men in all nations are at present laboring to produce."[92]

The Nay-Saying *Naturmensch*

Most criticism of Ortega's depiction of mass man has to do with Ortega's distinction between this type and noble man, its anti-thesis. It is correct to refer to noble man as the anti-thesis of mass man for the simple reason that Ortega views mass man's mindset and

92. C.S. Lewis. *The Abolition of Man*. New York: Macmillan Publishing Co., Inc., 1978, p. 86.

values as constituting the genuine essence of mankind: *Naturmensch*. Nobility is the recognition of its own shortcomings without taking recourse in emotional and psychological complexes. Hence to achieve any degree of transcendence we must be willing to recognize a clear and plainer state of being that we do not presently enjoy. The instinct of nobility is always geared toward the least common denominator. If there is a synthesis in this process, then it can be witnessed in the recognition and admission that the noble man makes to itself regarding his need to live in the world. Steeped in the world of things, events and other people, the noble man learns to seek shelter within itself. Thus the actions of noble man are informed by conquering a higher moral ground. Never losing sight of the inherent difficulties of this task, this makes the noble man more secure in his own abilities to attain transcendence, while remaining equally respectful of the insecurities fostered by life itself.

Ortega measures the true dignity and heroism of noble man in his ability to comprehend that life and all its exigencies, as Camus has lucidly pointed out, are meagerly absurd at best. Noble man does not turn this valuable understanding into social/political self-congratulatory verbiage. Neither does noble man rebel against himself, as is commonly practiced by those who have given themselves over to full-fledged inauthentic forms of life. Few today can honestly deny that the twentieth century, especially the post World War II era is an age ruled by mass movements, cults, cults of personality where people trade in their better sense to "belong" and social/political movements that do not demand a bettering of individuals, but only allegiance to causes and mother ideology. This primitive, herd mentality might be something that we do not necessarily may want to acknowledge, yet which serves as the most effective way to wholly embrace the mass man mentality without recourse to guilt or a sense of personal loss.

Having mentioned how most often commentators zero in on the differences between these two types of human being, we must not lose sight of just what is meant by the mass mind. Ortega's notion of mass man is reminiscent of the boy who, once summoned home, leaves the ball field and takes the only baseball bat with him. When beckoned to leave it behind, he merely utters that if he can't play, then why should anyone else. Now, to state this in such a blunt manner may prove embarrassing to discerning adults. But this begs the question of just how this attitude informs some lives — in many respects this attitude makes its appearance blunter, e.g. calumny, envy, bad will, disrespect for the autonomy of others — and the inhumanity of ideologues.

Personal inadequacies are justified by the mass man by his perpetual desire to seek commonality with others. Ortega understands the mass man's need for union with like-minded others as a desire to escape from the blatant need to exercise one's autonomy. The mass mind resents nobility. We thus come to the understanding that what truly drives mass man is the spirit of negation. Shrewd as *The Revolt of the Masses* is, it should come as no surprise that this is a misunderstood work. This work does not take its cue from lazy social/political generalizations and abstractions; instead it is informed by Ortega's readiness to listen to reality itself. His reasoning is informed by a lifetime of travel, editing, public speaking and above all mingling with diverse types of people. Those who refer to Ortega as an elitist should be reminded that the distinctions that he makes between the two types of people cut across class lines. It is also important to recall that Ortega reserves a great deal of criticism of the mass man for the landed aristocracy and other "society" types. This type, too, despite (or because of) their social positions, can be referred to as mass man. Ortega makes no distinction between a socialite mass man and a communist ringleader. What marks the mass mind apart from others is his truancy from himself. A life diffused in the material world is a life that sees no reservoir of meaning in itself. Ortega does not think that the importance of this existential category can be left to the generalizing whim of the social scientist. In fact he points out that this is a dangerous assumption, social scientists, in most cases, Ortega argues, being prime examples of the lack of autonomy of inauthentic mass man.

Chapter 4. Toward a Celebration of Man's Achievements

Implicit in Ortega's regard for culture and civilization is his respect for a genuine humanism. In many ways his is a humanism that is inspired by ancient and Renaissance models. Part of his respect for man's achievements springs from his understanding of the importance of neutralizing corrosive and debilitating cynicism. Ortega argues that man is a being that possesses an unquenchable desire for truth and understanding. However, as universal as this statement may sound, he remains cautious of the destructive forces that refuse to celebrate man's achievements. Ortega's humanism is shrewd enough to remain open to the understanding that human existence is in essence a perpetual target of cosmic — and worldly — resistance. And fundamental to this understanding is the notion that man must confront this resistance on his own. The aforementioned are two fundamental philosophical concerns that unfortunately have been all but eviscerated from philosophical reflection today. This is yet another reason why Ortega's brand of humanism is currently out of favor.

Ortega's cultural and philosophical hunger enabled him to pay close attention to cultural developments and technological innovations and inventions. His birth date, 1883, was a rather strategic time to witness the birth and subsequent development of the automobile, the steel beam, the all-glass building, the airplane and the start of the space age. In politics, he witnessed two world wars, the rise of Communism, Nazism, the start of the "cold" war and the proliferation of the techniques of the terror state. He also witnessed the explosion of cinema as a form of entertainment. In literature, he experienced the birth of high modernism and the subsequent derision and suspension of a great many literary and cultural traditions. In music, Ortega also witnessed the abandonment of orthodoxy for asymmetrical Serialism.

Ortega's generation not only saw the birth and expansion of these cultural and technological phenomena, but he also experienced the time that preceded them. This allowed for a point of reference, of comparison. Thus, as a philosopher he could easily appreciate the significance of the universal principles that he was witnessing. As a visionary thinker, he extrapolated the possible conditions that these movements would eventually bring about — their fruition and their exhaustion. What is so significant about this, however, is that while Ortega understands the technical aspect of these new developments, he also recognizes that in the impetus and strength behind any human endeavor there is found the vision and integrity of individuals.

As one who is concerned with the plight of culture and civilization, Ortega cannot help but to pay attention to the determining factors that produce, maintain or destroy the aforementioned. Because he lived in the midst of some of the fundamental developments of the twentieth century, Ortega was privy to the effort that goes into launching the technological progress that we enjoy today. This is evident in Ortega's essay "Medio siglo de Filosofia" ("Half a Century of Philosophy"), where he demonstrates the importance of his historical/biographical interpretation of human history as vital and "live." This "having-lived" first-hand knowledge creates an aura of understanding in people who are privy to such intricate developments, and who cannot easily make sense of the devil-may-care decadence of subsequent generations. Given the respect that Ortega has for machinery and technology, culture, art and literature, it becomes easy to understand his signaling out in *The Revolt of the Masses* the worrisome attitude concerning those who take progress for granted.

Beyond Ortega's concern for science, especially what he calls technicism, is his attention to the metaphysical essences that inform any of the achievements that he witnessed in his day. He asks: "Has any thought been given to the number of things that must remain active in men's souls in order that these may still continue to be 'men of science' in real truth?"[93] This is in keeping with the dangers of the aberrant sensualism that he warned about. What is left for man but to live outside himself once no substantial level of interiority is any longer cultivated? This negation of individual lived-vitality is another central point that Ortega makes in *The Revolt of the Masses.*

Ortega argues that because science brings about prosperity the masses should foster greater respect for science as the source of technicism, or what has come to be called applied science. In this regard, Ortega views science as having no cultural rivals given that neither politics, law, art, morality, or religion can take away its luster given the temporary bankruptcy of these other disciplines and institutions. He explains: "The monstrosity is increased a hundredfold by the fact that, as I have indicated, all the other vital principles, politics, law, art, morals, religion, are actually passing through a crisis, are at least temporarily bankrupt. Science alone is not bankrupt; rather does it every day pay out, with fabulous interest, all and more than it promises. It is, then, without a competitor; it is impos-

93. *The Revolt of the Masses*, p. 85.

sible to excuse the average man's disregard of it by considering him distracted from it by some other cultural enthusiam."[94]

Ortega's dread that the masses were seemingly getting used to living with "more" seems a natural response to a time of change. Present in his examination of this subject is a discrepancy between the appreciation of the new world that man created, with a great deal of effort, and the general response of the masses with an attitude of "things have always been this way." But a further point in Ortega's thought that has evaded the attention of many of his commentators is the slipshod mentality of "post-modernity." The mass man mentality, Ortega argues, has now triumphed to the degree that it has become institutionalized. Not the least important of his contentions is that even "this disregard of science as such appears, with possibly more evidence than elsewhere, in the mass of technicians themselves — doctors, engineers, etc., who are in the habit of exercising their profession in a state of mind identical in all essentials to that of the man who is content to use his motor-car or buy his tube of aspirin — without the slightest intimate solidarity with the future of science, of civilization."[95]

The problem as Ortega views it has to do with the disproportion of benefits that modern man receives from science and the level of gratitude and appreciation expressed by people today. Of course, the notion of gratitude does not entail the thanking of everyone responsible for bringing about any of our modern comforts. What is truly at stake in Ortega's commentary is the link between man's awe and wonder and his ability to be moved by his ability to control his material condition. This is a question of pathos. The inability to enact either of the two aforementioned poles is what Ortega refers to as barbarism.

But perhaps even a greater question that Ortega addresses is that of just how many creature comforts man can attain before the spirit that brings the latter to fruition begins to turn on itself.

94. Ibid., p. 86.

95. Ibid., p. 87.

While the cave painters of Altamira and Lascaux may not have been philosophers, they did exhibit a concern with transcendence. We can take transcendence here to mean the gulf that exits between man's desire to hunt a given animal and his inability to achieve this task. This serves as a fine example of the existential insecurity that concerned Ortega. This need to secure his own survival forced the "noble man" type to outdo himself in an attempt to succeed in the hunt. But more importantly this struggle also created an ethos that was commensurate with greater achievements still to come.

Ortega's Ideas about Science

Undoubtedly, Ortega has a great deal to say about science in *The Revolt of the Masses*. But most importantly, he offers strong insights into the relationship between an existential view of human reality and man's inventions. But as is typical of Ortega's subtle argumentative style, this notion is diffused throughout this work. For instance, in the chapter titled "Primitivism and History" he makes it quite clear that unencumbered nature alone does not give rise to civilization.

Nature, he tells us, is perpetually trying to "spring up anew" where civilization now exists or once existed. The tension that is present between civilization and nature is not only necessary, but also indicative of man's abilities.[96] This natural tension is a vivid metaphor that enables Ortega to explain the interplay between objectifying, material forces that rob us of existential authenticity and the exercise of personal autonomy. "Civilization is not 'just there,' it is not self-supporting. It is artificial and requires the artist or the artisan. If you want to make use of the advantages of civilization,

96. See: Dagobert D. Runes. *Dictionary of Philosophy*. New York: Philosophical Library, 1960. *Techne*: "The set of principles, or rational method, involved in the production of an object or the accomplishment of an end; the knowledge of such principles or method; art. *Techne* resembles *episteme* in the implying knowledge of principles, but differs in that its aim is making or doing, not disinterested understanding," p. 314.

but are not prepared to concern yourself with the upholding of civilization — you are done."[97]

Ortega's embrace as well as critique of science follows a traditional, rationalist notion of citing the limits of reason. His is not a "post-modernist" attack on science and reason by any stretch of the imagination. His measured and well-balanced understanding of the scientific enterprise places science and civilization as the logical culmination of man's understanding of his world. He cites the myopia of romantics, in their one-sided critique of science as a case of irresponsibility and negligence. The key to comprehending this angle of Ortega's work might be found in the meaning of his notion: "When your good romantic catches sight of a building, the first thing his eyes seek is the yellow hedge-mustard [a wild flower] on cornice and roof. This proclaims, that in the long run, everything is earth, that the jungle springs up everywhere anew."[98] Key to this statement is Ortega's not so subtle articulation that what affects the existential constitution of man the most is measured not in the long, but rather the short run. The human condition, that is, that portion that is assigned to each one of us is understood, embraced or neglected by the respective existential demands that it makes on us.

Implicit in his understanding of time as a cosmic phenomenon that man must contend with, is also Ortega's realization that human existence is lived on vital, individual terms, not with the possible security promised by scientific induction. Even if the face of our greatest creations eventually become tainted by time, we are nonetheless left with the existential conundrum of "what to do next?" the former being a form of fatalism which Ortega refutes, given his notion of human existence as a perpetual having-to-do.

Ortega's suggestion, however, in this regard is rather instructive. The failure of romanticism, Ortega suggests, lies in the fact that it is very easy for romantics to pass up vital necessity simply by basing

97. *The Revolt of the Masses*, p. 88.

98. Ibid., p. 89.

their assumptions on the force of future probabilities. This, then, seems to be the crux of *The Revolt of the Masses*: our willingness to negate our vital, existential individuality in exchange for the common place.

The mass man, Ortega tells us, necessarily assumes that the civilization that it has been born into — that has been handed down — has in fact come about by natural grace. Citing this myopic view, Ortega goes on to cite the sources of such a perspective. His grasp of this subject has us understand that the world of the mass man is ipso facto that of the natural order. By this we can understand that mass man places the inventions, culture and civilization that he witnesses around him in the same light as the cave dweller does with his natural surrounding. What can be more natural than to take for granted the customs (*vigencias*) and culture that we are born into? To break this destructive mold two essential elements must be met: first, the desire for protecting the status quo, and secondly the desire to create it in the first place.

It is interesting to view how Ortega's fine grasp of culture and science is directly tied to existential questions. At this point, it seems appropriate to remember that Ortega's notion of the forest in *Meditations on Quixote* equated the totality that is the forest to an empirical understanding of human reality that is best manifested as a vague sensualism. In that work, where he sets up his phenomenological understanding of human existence, Ortega asks the reader the simple question, what is a forest? The practical answer that immediately comes to mind: a tract of land of a given size and make up that sustains a cover of trees and undergrowth.

Formulated in this manner, the question takes on a rhetorical flair that has nothing to add to Ortega's existential concern. What does a forest sig- nify for one who enters it? One practical consideration is to know the way out. But even in this seemingly mundane way of navigating through a forest, that is — that substantial tract of land that is peopled by trees — is already to understand that there is always a subject involved.

Ortega's regard for the principles that rule over the civilized world serves to reiterate the contrast that exists between the forest, or raw nature, and the subject for whom it exists. He reintroduces this metaphor in *The Revolt of The Masses* because there he likens nature to the natural abode of primitive man. Civilization, on the other hand, frees us from the constraints of this primitive servitude. However, to construct cultural values, and most importantly to maintain them there must be an existential level of engagement that pays allegiance to the resistance that nature offers human reality.

It is virtually impossible today to deny Ortega's correct overall prognosis of the *status quo* of culture in the west. When he argued in the late 1920s that man cannot remain in nature's cave, if we are to have any illusion of living in the modern world, he does so from the understanding that science and technology would subsequently become more complicated and engaging.

Ortega develops striking parallels between the future complexity of human life and the advancement of science. At the individual level of self-conscious understanding of human existence demands more of itself than the biological drive can supply. This existential hunger and inquietude is an overflow of man's self-reflective creative ability. Of course, the creative act does not have to be self-conscious — certainly not all of the time — in order to seek some form of objectification. However, this is the case because man is a being capable of existential understanding.

But at the level of culture and civilization, Ortega argues, the conditions of life improve in proportion to the conditions that make material progress possible in the first place. His clear logic is indicative of a thinker who understands the plight of man's existential condition. He explains:

> Life gets gradually better, but evidently also gradually more complicated. Of course, as problems become more complex, the

> means of solving them also become more perfect. But each new generation must master these perfected means.[99]

Much too often today, these same questions that Ortega undertakes are treated sociologically as if they exist in an existential vacuum. Reading many of these vacuous, ideological accounts, one quickly gathers the hardened impression that the people who present human problems in such one-dimensional and unsophisticated ways are either ignorant of human history and man's existential inquietude or they are merely interested in ideology, not philosophy proper. Judging from the explosion of radical, anti-humanist, and philosophically and scientifically irresponsible fodder, one must conclude that both of the aforementioned conditions are very much at work.

Bad science is eventually undone by good science. What is unconscionable is the propagation of bad and irresponsible pseudo science for ideological reasons. A close reading of *The Revolt of the Masses* and other of Ortega's work reveals what in other times was considered a measured and rational respect for the limits of science and reason, limits that without a doubt are commensurate with man's limited understanding. Yet what is intellectually dishonest in the current fashionable assault on science is the myopic arrogance that attempts to secure "perfect" social/political conditions that protect man. Missing in this ideological milieu is the understanding that these social/political conditions are to be met by the same "limited" agents who engage in what ideologues consider non-objective science.

Ortega's reason works within the age old parameters that posit mind and reason as indispensable tools for human well-being. Again, Ortega's sharp reason has to do with his view of science and mathematics as objective in scope, even though he views these as pertaining to pure reason. These are forms of life and as such they are only one tool in man's epistemological arsenal. Nowhere does Ortega argue that science is non-objective. Whatever the motivat-

99. Ibid., p. 90.

ing factors in what is essentially a bandwagon, "post-modern" assault on science, Ortega's thought manages to remain classical in this respect.[100]

Ortega continues: "Historical knowledge is a technique of the first order to preserve and continue a civilization already advanced. Not that it affords positive solutions to the new aspect of vital conditions — life is always different from what it was — but that it prevents us committing the ingenious mistakes of other times.[101]

Ortega's philosophical acumen is nowhere more in evidence than in his anticipation of the "post-modern" form of nihilism that systematically negates history. When he argues that in lieu of the continual difficulties of human existence –the past should serve as a vital parameter — he does so from the understanding that to destroy the past, we must readily replace it with something greater. The vital importance of the past, Ortega observes, is that it serves us in the capacity to prognosticate of the future. Ortega equates knowledge of history with a vigorous ability for self-understanding. He alludes that both Communism and Fascism are retrograde mass movements that are fueled by "men who are mediocrities, improvised, devoid of a long memory and a "historic conscience."[102] For this reason Ortega views the past as somewhat of an inextinguishable candle. The proper thing to do, he tells us, is to "live at the height of the times," with an exaggerated consciousness of the historical circumstances.

However, Ortega is not wholly optimistic in his assessment of Europe. Every act of remembrance comes with a greater price than most people are prepared to pay. Decidedly, the height of the times

100. See: *The Sokal Hoax: The Sham that Shook the Academy*. Lincoln: University of Nebraska Press, 200. This article is a fine example of the inherent desire of "post-modernists" to undermine and "deconstruct" any tenements of the objective world. Sokal presents a clear exposition of the ideological motivations of the proponents of "cultural studies" at the expense of science and vital culture. Being ideologically motivated, these attacks are certainly not grounded in an awe and genuine desire for understanding.

101. *The Revolt of the Masses*, p. 91.

102. Ibid., p. 92.

belongs to an empty chorus that he refers to as mass man. As remembrance continues to become supplanted by self-imposed ignorance, the triumph of mass man also becomes more inevitable.

In *Meditación de la Técnica (Meditation on Technique)*, Ortega offers much more than an analysis of the nature of technology and scientific know-how. This work follows the trajectory of his meta-ontological analysis of man. In short, most of Ortega's works can be classified as metaphysical anthropology. He begins *Meditación de la Técnica* by negating the behaviorists notion that posits instinct as the strongest answer to the question, "why does man live?" In its place Ortega cites reason and intuition as the tools that guide life. Thus, he answers the question by arguing that man is the only being that possesses the ontological/metaphysical capacity to take its own life.

Much like is the case in *Meditation on Hunting*, where the decisive emphasis is placed on man in the cosmos; *Meditation on Technique* is equally a misleading text on man in a technological age. Present within any of Ortega's analysis is an implicit regard for man's existential/metaphysical condition and not with an isolated view to science and technology.

In *Meditations on Technique*, for instance, he makes it clear that technology exists as an outflow of man's vital energy. It is important to recognize that Ortega does not equate human life with the merely biological. When he argues that man's life doe not "coincide with the profile of his organic needs" he does so from the understanding that man has the capacity to manipulate his environment, which incidentally includes his biological life.[103]

Instead what is said to fuel this manipulation is what Ortega views as the difference between biological drives in animals and needs in man. While the former is felt as an instinctual drive that the animal cannot control, and which when not met will prove fatal, man instead feels the need to surpass the merely biological.

103. *Meditación de la Técnica*. Madrid: Colección Austral, 1965 [my translation], p. 19.

This understanding is essential to understanding *The Revolt of the Masses*. There is in Ortega, as is the case with other thinkers, a great deal of ineffective and confused hermeneutics. When we base our understanding on research, on reading an author based on what he actually said and avoid applying the use of extraneous "theory" then we find that what any interlocutor has to say should not necessarily become the center of attention. Careful attention ought to be paid to the realization that hermeneutics cannot be a word-game that respects no boundaries, much like a balloon that can be inflated at will without consequences.

Some commentators have taken *The Revolt of the Masses* out of context and used it to soothe some extrinsic need. What motivation is at work in the cases of these egregious misrepresentation — of any texts — we can only speculate, but we can rest assured that at least two of these can be easily identified: intellectual laziness and superficial analysis.

What informs and motivates *The Revolt of the Masses* is nothing less than a philosophical analysis of man, how he fits in the scheme of a greater reality, and how we participate in such a reality. What we must avoid at all cost is the vacuous speculation that asserts that Ortega is engaged in social/political "theorizing." This, however, is a difficult habit to break, especially in an age when metaphysics has been subsumed by positivistic sociology, and when even private existential concerns cannot escape the "scrutiny" of politicization.

A fine case in point is Ortega's placing man outside of nature, as pertaining to a phenomenal cosmic entity. When he discusses the differences between mass man and noble man, he is entertaining the argument that man's creations do not come about through some vague evolutionary tract that delivers the goods we enjoy ready made. And if it the case that man's social world revolves around the admission of conviviality, then Ortega poses the further question, what parties are responsible for cultural creation and its upkeep?

Inherent in his understanding of this problem is the notion that some men desire to remain in the cave of servitude to nature.

What makes man a phenomenal cosmic entity is his ability to view his primal circumstances as literally that: a reality that only appears to be "this or that" to itself. This is no less than the discovery of man as transcendence. How this discovery impacts humanism, art, science, psychology, etc., is precisely where *The Revolt of the Masses* makes its greatest contribution.

Thus technique, as Ortega defines it, modifies or reforms nature and/or our circumstances. Technique is: "Man's reform of nature in light of the latter's lack of fulfillment of our being."[104] Hence human existence, when viewed as other than biological, is embraced as transcendence. But transcendence is neither easily attained or is it transparent in make up. Transcendence manifests itself as resistance, thus Ortega's notion of nobility. We find that there are a number of difficulties involved in our acts of transcendence. In the process of refining our existence, of making our condition easier, we are greeted by immense obstacles and setbacks. Ortega writes:

> Technique is the effort to save us effort, in other words, it is what we do to avoid, in its totality or partially, having-to-do (*que-hacer*) what circumstances necessitate.[105]

Ortega's measured critique of science and the limits of reason are possible precisely because he is able to balance it with an in-depth look at the origin of science. As such, science is the repository of man's transcendence from the biological servitude to nature and his ability to control his destiny as best as he is able. *Meditación de la Técnica* is essentially Ortega's major essay on civilization and its origins. All human activity, he advises, serves as a mode of technique, which places in motion the mechanisms to save our circumstances. To this assessment he adds: "If our existence was not from the beginning one forced to construct with the materials found in nature, the ex-

104. Ibid., my translation, p. 21.

105. Ibid., my translation, p. 35.

tra natural aspiration that man is, none of our techniques would be possible."[106]

The implication of *The Revolt of the Masses* that "natural man" is assaulted by nature and makes nothing of it serves as an essentially description of mass man. While it is the noble man that is pestered and even paralyzed by technical and existential concerns, the mass man remains oblivious to such possibilities. But because the civilization that is created is enjoyed by all, for this same reason Ortega argues that man's creations are in constant peril of disappearing. In a rather paradoxical twist, this confirms the truth of what Luddites assert as the contingency of a linear model of civilization. Ortega too argues that this is the case even though for drastically different reasons. While Luddites envision a world stripped of technological advancement, Ortega makes the case that this is very much a real possibility given a certain level of creative/existential complacency that is reached by many while enmeshed in modern comforts. This possibility exists precisely because of the "relaxed" attitude that cannot help but to take for granted that which does not come naturally. To live, then, is to "feel the subsoil of history" beneath our feet. This is the ideal. But the reality that Ortega argues for is that today it is man who has failed and not science or technicism.

106. Ibid., p. 48.

Chapter 5. Toward an Aesthetics of Life

Ortega's preoccupation with the vital, that is, existential aspects of human existence and the objectifying external forces that such a life must contend with is in keeping with the spirit of philosophy. Ortega retains the ancient sense of awe that is so essential to ancient philosophy. Perhaps somewhat due to temperament, his work never strays too far from his concern for man as the inexhaustible topic of reflection. Part of the theme that he develops in *The Revolt of the Masses* has to do with the biographical nature of man.

Ortega's work never disappoints in its humanistic treatment of man. His balancing of science, technology, the social/political and the existential succeeds because he understands that man cannot betray his existential/metaphysical condition. Consider his short essay "En la muerte de Unamuno" ("On Unamuno's Death") and how he evokes the same existential gravitas that informs all of his work. Ortega explains how on the first night of 1937 he received a telephone call informing him about Unamuno's death. He goes on to compare the vital vicissitudes that inform Unamuno's life and the pathos for life that Unamuno embraced and which dominated his thought and discourse.

What we find in Ortega's analysis of Unamuno's character is a decisive and vital sense of conviction that allows him to follow an unwavering life-path. Ortega admires Unamuno's courage and discipline. He describes Unamuno as an individualist who having reflected on a great number of existential themes finds solitude a suitable companion. According to Ortega, what Unamuno demanded of himself: sincerity, vital convictions and a felt sense of life's stages, he also demands of others with whom he enters into conversation. Ortega explains this:

> Unamuno is now united with death, his perennial friend/enemy. All of his philosophy has been, like Spinoza's *a meditatio mortis*."[107]

Ortega goes on to say that Unamuno forced Europeans to reflect on the seriousness of life during a time when they had become distracted from the essential human vocation of having to die by concentrating on mundane things. Ortega refers to mundane things as "the things that reside with life."

As a philosophical theme, our existential sense of purpose quickly gets muddled with the mundane things of the world. As such this feeling for life becomes trampled by more practical considerations. Critics of both Ortega and Unamuno and other thinkers that follow an existential path will guff with smugness at the thought that large portions, or even all of life ought not to be taken too seriously. This point, too, merits consideration in that the truth that it proposes can easily be verified. Yet this is not Ortega's point. Both Ortega and Unamuno readily agree with such critics in their arguing for individual vitality, as this became consumed by an overbearing rationalism and ideology in the twentieth century.

The point about the seriousness of life that Ortega makes throughout his work is that what is serious is precisely that which is felt, lived and not what is merely arrived at through extraneous instruction. This, of course, is where vocation makes its greatest contribution to his work.

107. *The Revolt of the Masses*, p. 85.

Another manifestation of this Ortegan notion of seriousness of purpose, of individual character is found in an essay titled "El derecho a la continiudad," where Ortega treats the idea of personal discipline in relation to the overall effect that this has on the nation. This question has to do with cultural continuity, especially as this evinces a tranquil and prosperous peace. This essay, even though short, is an interesting look at personal discipline, cynicism and the overall health of the state — concerns which we ought not to forget, are the driving force behind *The Revolt of the Masses*. Here Ortega writes: "So many states, men and things have failed us that we have lost our intimate right to trust anything any longer.[108]

Midway through *The Revolt of the Masses* we encounter a chapter titled: "The Self-Satisfied Age." This seems appropriate because Ortega moves from a historical analysis of the masses, their expansion into all areas where before they were led by minorities, and into an analysis of just what he means by mass and noble man.

The Revolt of the Masses manages to find a resolute balance between the despotism of the higher and the vulgarity of lower classes by ignoring the line that has traditionally separated the two. Mass man is found in both of these classes. Ortega is quick to point out that the mass mind constitutes a despotism and vulgarity of perspective that is irrespective of social standing. Mass man makes up the bulk of both of these traditional enclaves. This is one respect where Ortega's thought proves to be very much ahead of its time. The idea of meritocracy, he argues, has never been taken too seriously by most scholars, especially in liberal democracies where it is de facto at work, precisely because its greatest obstacle is the interference of mass man. Another reason for this has to do with the question of social/political "theories" that completely ignore the inherent differences in vocation, temperament; existential, visionary, spiritual, moral and intellectual that informs human reality. What these social/political theories have so effectively promoted is the vacuous, dead end notion of "perpetual revolution." Having understood this

108. Ibid., p. 261.

perfectly well, Ortega attempts to go beyond this historical *cul de sac* by describing the conditions that inform the man of flesh and blood and not romantic theoretical projections. While the timeless questions that Ortega addresses might come across as "sensitive" points with many today, his is a sincere philosophical voice that cuts through theoretical and insular case studies that opt for tired and weary "solutions."

The mass mind that fuels the self-satisfied age does so out of the belief that it possesses the right, and not just the possibility of ruling regardless of the capacity to do so. It is worth citing the entire passage where Ortega explains what he considers to be the three main psychological characteristics of the mass mind.[109]

This is a very interesting analysis because Ortega makes use of the word "inborn," an A-priori reckoning with human reality that is out of fashion in today's modish intellectual milieu. By suggesting that people may be predisposed to some given view of reality, Ortega, who in other works has argued for life as having-to-do (*quehacer*) leaves open the possibility of a hierarchical epistemology. This is very much in keeping with his early ideas on Perspectivism. Ortega argues that the dominant principles that rule over human reality and truth are objective, but that all individuals must encounter these principles of their own accord: Perspectivism. However, this does not mean that all views are equal, only that objective reality makes itself known in relation to the degree that individuals are willing to engage it.

---

109. *The Revolt of the Masses*. Ortega adds: "(1) An inborn, root-impression that life is easy, plentiful, without any grave limitations; consequently, each average man finds within himself a sensation of power and triumph which, (2) invites him to stand up for himself as he is, to look upon his moral and intellectual endowment as excellent, complete. This contentment with himself leads him to shut himself off from any external court of appeal; not to listen, not to submit his opinions to judgment, not to consider others' existence. His intimate feeling of power urges him always to exercise predominance. He will act then as if he and his like were the only beings existing in the world; and, consequently, (3) will intervene in all matters, imposing his own vulgar views without respect or regard for others, without limit or reserve that is to say, in accordance with a system of "direct action," p. 97.

An even greater problem that Ortega points out in the cyclical nature of human history is our understanding or failure to recognize human existence as always offering resistance to our individual lives. He tells us, "all life is the struggle, the effort to be itself."[110] Human life, then, is best understood as resistance. This resistance is present in every one of our endeavors, trivial as they may seem at times. But resistance in itself, Ortega argues, is the manifestation of life in its human form. Thus forces us to discover the self and our place in the scheme of things: my destiny or vocation. The struggle to become what my intuition informs me that I am is part of the struggle to be. To live, then, is to confront and develop ourselves. This is very much in keeping with the existential notion of life as project. Hence in Ortega's view, to posit life as resistance, as other thinkers like Dilthey and Heidegger have also argued, brings about two central realizations: humility and a sense of awe. These two conditions, when maintained throughout the course of one's life become fundamental ingredients in not following the somnambulist sub-existence of the mass man.

However, this condition is constantly tested and castigated by the cyclical nature of reality proper throughout the course of our lives because resistance acts as a force that keeps us from degenerating into complacency. Ortega writes: "A world superabundant in possibilities automatically produces deformities, vicious types of human life, which may be brought under the general class, the 'heir-man,' of which the 'autocrat' is only one particular case, the spoiled child another, and the mass-man of our time, more fully, more radically, a third."[111] For this reason Ortega argues that all forms of social/political "revolution" only manage to create another mass man ruling class where nothing can change. This is a futile cycle.

The undercurrent of *The Revolt of the Masses* never strays from the existential/metaphysical essences that inform human existence. For this reason Ortega argues that the best antidote against the compla-

110. Ibid., p. 99.

111. Ibid., p. 100.

cency of the mass mind is one of sincere, personal renewal. "I am I and my circumstances" and "man as a cosmic phenomena," are the hallmark of individual reflection.

While "post-modern" ideological critics scuff at the idea of material progress, culture and civilization, Ortega merely argues from the perspective of a guardian of the former. Rather than offering ideological invective and a "theoretical" diatribe, his sincere reflection recognizes that everyone benefits from the technology and science of civilization. Ortega's analysis of the status quo of science and civilization essentially came about from the understanding that civilization, as we have come to know it, is a fragile enterprise. His was also a preoccupation with what he noticed was the sweeping tide of nihilism that Nietzsche warned about — a time of debauchery and "games" rule the day.[112]

Ortega's characterization of mass man as one steeped in sensuality and trivialities is prophetic of "post-modern" disregard for seriousness. He writes: "His propensity to make out of games and sports the central occupation of his life; the cult of the body — hygienic regime and attention to dress."[113]

The problem of the aforementioned, as Ortega sees it, is not so much that sport and games or dress for that matter are negative things in their own right. In *Meditations in Hunting* he makes it clear that what validates hunting, as an ancient sport is precisely the

112. Nietzsche, Friedrich. *Beyond Good and Evil.* Translated by Walter Kaufmann. New York: Vintage Books, 1966. "Today, conversely, when only the herd animal receives and dispenses honors in Europe, when "equality of rights" could all too easily be changed into equality in violating rights — I mean, into a common war on all that is rare, strange, privileged, the higher man, the higher soul, the higher duty, the higher responsibility, and the abundance of creative power and masterfulness – today the concept of greatness entails being noble, wanting to be oneself, being able to be different, standing alone and having to live independently. And the philosopher will betray something of his own ideal when he posits: 'He shall be greatest who can be loneliest, the most concealed, the most deviant, the human being beyond good and evil, the master of his virtues, he that is overrich in will. Precisely this shall be called greatness: being capable of being as manifold as whole, as simple as full.' And to ask it once more: today — is greatness possible?" p. 138

113. *The Revolt of the Masses*, p. 100.

mutual respect and union of man and nature. The self-respecting hunter, he goes on to argue is always a conservationist first. The major reservation that his thought has for what he considered the vacuous regard for sport and the cult of personality and the body has to do with these activities becoming ends in themselves. The apparent dichotomy that he presents is not one between the fulfilled life and that of sport, but rather the notion that a fulfilled life can incorporate games into its fullness. This trivial and superficial utility of mass minded man is a direct consequence of the nineteenth century, he contends. This "allows the average man to take his place in a world of superabundance, of which he perceives only the lavishness of the means at his disposal, nothing of the pain involved."[114] But rather than offering a destructive critique of civilization that does not contain solutions, Ortega instead argues for an ongoing corrective of the possible disproportion between strife and its achievements.

The "contact with the substance of life" that Ortega argues for bespeaks a great deal about conditions that inform modern life. Where is the elitism, as his critics suggest, in arguing for a greater appreciation of the material conditions that we currently enjoy? The degree of sophistication and existential self-awareness that Ortega seeks demands that life be turned into a poetic juncture. What he essentially promotes is a new aestheticism of life itself. This aestheticism is a recovering of the philosophical awe that has always served as the driving force behind human creation. What astonishes Ortega most is the fact that the greater the level of comfort and the rising of the standard of living, the less that most people seem awed by such things. Embedded in the more obvious themes of *The Revolt of the Masses* one finds other, more latent, themes that are vital to our understanding of the world today. Ortega argues that to take up such an existential stance toward the questions that life presents us with is also to create the necessary resistance to stupidity and banality that always threatens to keep us malcontents.

114. Ibid., p. 101.

The self-satisfied man, he tells us, thinks in terms of finite units of time that are just long enough to soothe his biological urges. The major characteristic of this mindset is the idea that nothing is permanent, and thus we ought only concern ourselves with the present. At an existential level, the operating model that this represents is one that embraces the killing of time — graver concerns are only mended for the time being. Thus the self-satisfied man is the direct opposite of the noble man. While the noble man adjusts to reality — to what can and cannot be — the self-satisfied man embraces a life of falsity, only paying lip service to what cannot be while holding steadfast to failed views and opinions. "The mass-man will not plant his foot on the immovably firm ground of his destiny, he prefers a fictitious existence suspended in air....Hardly anyone offers any resistance to the superficial whirlwinds that arise in art, in ideas, in politics, or in social usages," he goes on to write.[115]

The greatest failure of the self-satisfied man is his inability to understand and thus come to terms with his own existence. What this type of man fails to understand most is that human existence is fluid, and that the endless array of problems that it presents will continue to assault us as long as we live. Life as having-to-do (*quehacer*) serves as the central existential category that informs *The Revolt of the Masses.* This having-to-do rules the life of noble man because he embraces difficulties not necessarily for his enjoyment, but from a deeply rooted understanding of reality. This is not the case for the mass man. The mass man, on the other hand, deals with difficulties in two ways: either reluctantly and thus cuts corners in every aspect of his life, or he is completely oblivious to his destiny and the duty that he has embraced. Here Ortega cites ignorance, much like Parmenides' notion of not being able to know "what does not occur to us." Yet to view the question in such a way is to come full circle in our explanation of the differences between mass and noble man.

115. Ibid., p. 105.

However, if we take a materialist line of thought that stresses that our circumstances create thought, we are then still left with the difficulty of explaining what ingredients must be contained in the circumstances for ideas to dawn. This materialist interpretation supplies no knowledge as to what ingredients must be present in the formation of ideas. What remains latent in our circumstances — in short — for us to recognize? For instance, the numerical value that represents the gravitational pull of Earth is the same today as it was five thousand years ago. By the same token this numerical reality is also present for the infant in the crib, as it is for primitive people and astrophysicists. Perhaps the ability to perceive is nowhere better witnessed than in sports where, even amongst professionals, the respective abilities of the players demonstrates a broad range.

Ortega offers an analysis of perception that recognizes that, while we may not be able to answer this riddle in great detail, he does suggest that the importance of our circumstances lies in awaking man's innate capabilities. He addresses this question in his book *What is Philosophy?* by arguing,

> So that thought may act, there must be a problem before it, and in order that there must be data. Unless something is given us it does occur to us to think of it or about it; and if everything were given us we would have no reason to think.[116]

Fortunately, this epistemological conundrum does not serve as an impediment to Ortega's existential/metaphysical rendition of man in *The Revolt of the Masses*. Part of the reason for this has to do with Ortega's emphasis on the life-conditions that individuals make for themselves. Hence part of the question of human circumstances, according to Ortega, already involves how we relate to ourselves existentially. How we understand the limitations of our bodies, our emotional/intellectual capabilities and our role in the space-time continuum are all central components that guide us in embracing our circumstances.

116. *What is Philosophy*. Translated by Mildred Adams. New York: W.W. Norton & Company, 1964, p. 134.

The middle ground that makes this question a much more practical and vital dilemma than some commentators realize has to do with man's ability to recognize his ignorance. This implies that whatever we do know always raises a red flag signaling that which transcends it. This innate categorical mechanism is precisely what Ortega cites as the central component of social life. Based on what we are able to "see", we can come to a greater understanding of our inabilities, limitations and vocation. This requires a degree of sincerity and good will, as it were, that advises us on what course to embark on. Failure to grasp this reality becomes for Ortega the initial distinction between mass and noble man. He adds: "It is knowing that we do not know enough, it is knowing that we are ignorant. And such, strictly speaking was the deep meaning of 'knowing that he did not know' which Socrates attributed to himself as his only pride."[117]

Even the self-satisfied man, then, knows something about the privileges that he enjoys. He knows that these are inherently difficult to bring to fruition and easily destroyed. Again, the question also hinges on good will. In some of Ortega's most brilliant and insightful expositions of the mass mind, he cites the example of the cynic as the best of all self-satisfied minds. The cynic is a parasite of civilization, sponging off the sweat of those who create while he himself creates nothing. He can take this line of action, Ortega tells us, because the cynic, while being sinister in his radicalism, is nonetheless not a blind fool. The cynic criticizes and destroys, Ortega goes on to argue, because he knows full well that he cannot easily undo the work of civilization. Ortega explains:

> The cynic pullulated at every corner, and in the highest places. This cynic did nothing but saboteur the civilization of the time. He was the nihilist of Hellenism. He created nothing, he made nothing. His role was to undo — or rather to attempt to undo, for he did not succeed in his purpose. The cynic, a parasite of

117. Ibid., p. 134.

> civilization, lives by denying it, for the very reason that he is convinced that it will not fail.[118]

As an exposition of some of the supreme forms of inauthenticity in human existence, *The Revolt of the Masses* presents us with difficult questions that transcend our clichéd, trite and modish way of speaking today. In this respect alone, Ortega's work is representative of the philosophical enterprise. It would seem rather superficial for us to take an argumentative line that only cites man's environment, his social political world as the dominating factors in man's revolt. Such an intellectual stance totally ignores man's internal dimension.

It would seem equally banal to view man, his psyche, emotions and works as proceeding solely from a materialist anti-humanism that stubbornly attributes the creation of the human world to blind natural cause.[119] But this is precisely what our many varied forms of positivism allege. Ortega's work is a steadfast rebuttal of all forms of materialism regardless of how ingenious its manifestations. What is especially daunting about philosophical thought today is that because this dominant positivism has waged a full scale assault on metaphysics, no manner of discourse is currently excepted that does not confirm to the many current forms of materialism. For this reason some commentators of *The Revolt of the Masses* find themselves at odds with its author, who assures the reader that the social/political material world is but a veiled semblance of human existence. What is real and will always remain so, Ortega explains repeatedly, is a cosmic entity whose life is more informed by his inherent abil-

118. *The Revolt of the Masses*. Ortega goes on to say, "none other could be the conduct of this type of man born into a too well-organised world, of which he perceives only the advantages and not the dangers. His surroundings spoil him, because they are 'civilisation' that is, a home, and the *fils de famille* feels nothing that impels him to abandon his mood of caprice, nothing which urges him to listen to outside counsels from those superior to himself. Still less anything which obliges him to make contact with the inexorable depths of his own destiny," p. 106

119. Every facet of Ortega's work can be considered a reaction to positivism in all of its guises: aesthetic, moral, historical and social/political.

ity or the lack thereof for autonomous self-knowledge. Instead, the more malicious and limited critics of this work and of Ortega as a thinker negate this reality. This pathological, myopic condition may be best explained as covering the sun with one finger.

Perhaps no greater essay on the nature of revolt has been written than Camus' *The Rebel*. Like Ortega, Camus has a strong insight in the recognition of man's existential component that simultaneously transcends and creates human history. While Ortega's depiction of the cynic as a proto-mass man places this entity against noble man, he also argues that this atrophied (inauthentic) existential conditions becomes manifested as a form of history that nobility must contend with.

Our current dearth of imagination in paying close attention to historical data — and the absence of good will that goes with it — cannot progress beyond a form of sterility that cannot help but to increase the power of the state. All forms of the modern state, Ortega makes clear, are nothing other than the institutionalizing of the inauthentic values of mass man. When Ortega asks, "Who rules in the world?" at the end of The Revolt of the Masses, he confirms what he has said all along about the mass man. At the end of the book he demonstrates that the social/political manifestations, like the damage caused by termites, may be present, unnoticed, long before being seen or felt. Camus also makes this clear:

> On this level, therefore, history alone offers no hope. It is not a source of values, but is still a source of nihilism. Can one, at least, create values in defiance of history, on the single level of a philosophy based on eternity? That comes to the same as ratifying historical injustice and the sufferings of man. To slander the world leads to nihilism defined by Nietzsche. Thought that is derived from history alone, like thought that rejects history completely, deprives man of the means and the reason for living. The former drives him to the extreme decadence of "why live?" the latter to "how live?"[120]

120. Albert Camus. *The Rebel*. Translated by Anthony Bower. New York: Vintage International, 1991, p. 249.

Ortega sees in The Revolt of the Masses a vacuous revolt, a form of nihilism that because it does not recognize anything of intrinsic value in human existence battles against itself. But this battle is a negative one — that is, it takes the form of a negation against the self. While history has traditionally been the struggle for the objectification of nobility, Ortega instead sees the "revolt" as history-as-moral turpitude.

As one reads *The Revolt of the Masses* — carefully — we are treated to a meticulous dissemination of the major categories that inform modern, or what is by some accounts "post-modern" life. The destruction of every one of these categories for the sake of further "liberation" leaves all regard for existential life a barren landscape. Under such conditions the metaphysical essences that inform the entity that is the human person is swept aside as a colossal example of what Sartre has called bad faith: we convince ourselves that by not glancing in the direction of cosmological truth, we eventually succeed in vanquishing it.

Chapter 6. Nihilism and Collective Banality

William Barrett begins his seminal work *Time of Need: Forms of Imagination in the Twentieth Century* with an exposition of modern nihilism and classical notion of tragedy. In juxtaposing these two he demonstrates a clear demarcation between a time when human reality made sense of human existence as a tragic comedy and that which consumes itself in nihilistic meaninglessness. Barrett's comparison is enlightening. The former is indicative of a stoic attitude toward life, and perhaps the latter can be explained as issuing from a lack of imagination.

Here it is appropriate to recall Ortega's idea in *The Revolt of the Masses* that revolt does not mean political revolt. This angle, unfortunately, seems to have solidified into the only viable possibility for man by the latter half of the twentieth century. Ortega's arguments do not speak to ideologues or those steeped in social/political maneuvering. This aspect of *The Revolt of the Masses* strikes a cord with readers who come to this book expecting to read a work of political philosophy.

The question of nihilism is not formulated in Ortega's work as a straight analysis of meaning and angst. Nowhere throughout his work does Ortega argue that man's revolt or what is the revolt of the masses issues from angst. On the contrary, he views modern man, especially man in the late nineteenth and twentieth century as privy to a cultural, spiritual and material historical reservoir that is unprecedented. If man, today, Ortega suggests, feels lost, this comes as the result of a failure to understand that life is an existential *quehacer* or perpetual (having-to-do).

Ortega explores this question by analyzing the relationship between man's existential inquietude — hence freedom — and the life supplied by material progress and technology. This, then, is one of the major crossroads that Ortega exposes in *The Revolt of the Masses.* Interestingly enough, today we cannot, with intellectual honesty, negate the reality of either of the two aforementioned conditions: the first because it is a vital differentiation of cosmic man, and the latter because it has infused man's freedom with seemingly infinite possibilities. The major problem, as Ortega sees it, having to do with the possibilities that material progress creates is precisely that of a negation of existential essence. The apparent tradeoff in this exchange seems to be fueled by the numbing of our existential inquietude. Again, let us reiterate, that not until we have understood the form of revolt that Ortega makes use of can we precede to an understanding of his many proactive considerations.

One fine example amongst many of this question can be found in *España Invertebrada* where Ortega argues — in a chapter titled "The Absence of the Best" — that the early part of the twentieth century saw Russia and Spain without a cultural infrastructure that would permit the minorities to enact improvements. About Spain, he has the following to say: "The autonomous personality that adopts a consciously individual attitude toward life, has been a rarity in our country."[121]

The interplay that exists between man as a cosmic being, and a social entity — given that part of Ortega's existential concern is that

121. José Ortega y Gasset. *España Invertebrada.* Madrid: Revista de Occidente, 1921, p. 134.

man co-exists with others — cannot easily be subsumed by social/political positivism. Instead, Ortega argues that the sole purpose of social/political organization is to safeguard man's metaphysical/existential possibilities. But even more importantly, the conditions imposed by the state are indicative of the pathos of its members.

In *Historical Reason* Ortega argues that neither, material or philosophical progress are the result of a linear history. The very possibility of a relapse of man's achievements indicates the degree of truth that inheres in a cosmic view of man as having-to-do. Ortega begins the first half of this work by addressing the Faculty of Philosophy and Letters of the University of Buenos Aires, in October of 1940 by saying that his topic included nothing less ambitious than everything having to do with the world of man. But having said that, he then goes on to say: "But we will treat all that in a manner befitting this hall. In this place, from this Chair of philosophy, we don't speak of things, but of what is 'essential' about them. Although the air is scent-free, this is a Chair of essences."[122]

The import of this pronouncement is felt through all of his work: history is created by individual, vital reason. If individual essences are responsible for what we call the social/political, then clearly Ortega suggests that the world we create cannot be separated from our existential vocation.

Critics of *The Revolt of the Masses* today, when the social/political climate is dramatically entrenched in the materialism that Ortega warned about, will, out of self-survival negate the importance and insightfulness of this work. Part of what Ortega depicted in *The Revolt of the Masses* is the domination of the democratic process by the irrational and illiberal forces of the mass mind. A comprehensive glance at the social/political world scene today, cannot help but prove Ortega's thesis correct.

More closely allied to this point, we must realize that Ortega's minority noble man — is the propagator of genuine democracy. The

122. *Historical Reason*. Translated by Philip W. Silver. New York: W.W. Norton & Company, 1984, p. 13.

failure of some critics to recognize Ortega's democratic impulse lies in the inability to tie in aristocratic — what is essentially a noble vocation and disposition — with the establishment and upkeep of democracy. Part of the evidence for this notion is that nobility, according to Ortega, does not impose itself on others. Nobility by its very disposition is self-contained in its propensity to live off its own ideas and moral convictions. Mass man, on the contrary, possesses no moral center and is ruled by the whim of popular opinion. Mass man is consumed by meaninglessness, a spiritual vacuum that must be continuously filled by the political. "The uneasiness of modern man" Barrett explains, "arises from a rupture between himself and nature that leaves him homeless within the universe."[123]

The question of nihilism and meaninglessness plays a gargantuan role in *The Revolt of the Masses*, in Ortega's dissection of man as a cosmic being. What is so essential about this predicament — to cite Ortega — is that truth is eviscerated by the former in exchange for immediate gratification. The mass man reaps the goods that the tide of history delivers to his doorstep. What goes unnoticed or is completely sidestepped is the inspiration for such history.

As an essay on the question of meaning, *The Revolt of the Masses* covers this question in a much more systematic manner than its author is readily credited with. Ortega addresses this concern from diverse angles. These include the question of what happens to people when they find themselves in the midst of massive societal structures.

> Ortega first follows the origin and trajectory of human agglomeration and attempts to dissect the plight of autonomous individuals imbedded within. Part of this dissection has to do with a metaphysical study of man's essence or animating force. There, Ortega tells us, we discover a pronounced difference amongst what informs different individuals, and how these qualitative differences become manifest in the public realm. These qualitative anomalies point us to the nature of human history and how some people enjoy the passage of time, as what Ortega refers to as a life of inertia, while others become consumed by time in their struggle to become differentiated.

123. William Barrett. *Time of Need: Forms of Imagination in the Twentieth Century*. New York: Harper & Row, Publishers, 1972, p. 371.

Implicit in *The Revolt of the Masses* is the realization that "post-modern" man has arrived at a self-consuming sensualism that recognizes itself to be a dead end while simultaneously finding an occasion to rejoice. Hence, a close reading of this work shows us how this is the case, not as one more "theory" but rather through citing concrete examples of this malaise. Ortega's analysis cuts through anthropological questions that ask what is man? — questions of the nature and future of technology, but most importantly he addresses concerns having to do with the spiritual/moral fiber of man. Perhaps one of the strongest impressions that *The Revolt of the Masses* makes is Ortega's suspicion that "post-modern" man is nothing more than a nihilistic, world-weary entity that suffers from a collective fugue state.

But meaninglessness for "post-modernity" does not necessarily have a paralyzing effect. Ortega demonstrates, and "post-modern" theorists confirm how today, meaninglessness becomes the very thing that we invoke. This unprecedented and pandemic meaninglessness is celebrated in more imaginative, or liberating ways than any angst and alienation found in the works of Kierkegaard, Nietzsche, Sartre or Camus. Ortega explains:

> Those who are younger will see, in future years, the extent to which it is not a matter of no concern to have to live together without any final appeals court. Because the threat that logic will go up in smoke and be just another human utopia, another vain illusion that has lasted twenty-five hundred years...means quite simply, that the very notion of truth is on the wane; and thus, in a real sense, if this threat is a fact and no remedy is found, there will cease to be true or false things, nothing will have its own, clear-cut being, and the light that truth represents in the primordial darkness of our lives will be extinguished and darkness will reign.[124]

Of course, one must differentiate between a crisis, where truth is merely undergoing an adjustment, and that of its flat out negation. Ortega addresses both these questions throughout his work. While the former he analyzes in his works on metaphysics, he takes

124. *Historical Reason*, p. 176.

the latter's effect as the starting point of *The Revolt of the Masses*. He adds:

> But the earthquake has occurred not only in scientific or theoretical reason, but also in practical reason, and at the same time. I shall skip morality, which, as the rational law of conduct, went up in smoke some forty years ago, leaving the weight of custom as the sole regulator of human behavior.[125]

The adoption of meaninglessness as a form of existence is a sign, Ortega goes on to argue that twentieth century life, especially that of the post World War II period is enthralled by a spurious rejoicing in the form of banalities. This is why Ortega argues in several places that man today lives for the moment and that all our cures are merely designed as temporary patch jobs. Ortega ties in the inability of the mass man to create or achieve anything substantial with what today we refer to "post-modern" theorizing. While the theorizers have decreed that there is no truth, that nothing can be known for certain, they continue the schizophrenia of fashioning an endless array of insipid theories. But why, we may ask? What is the point of theorizing, if we admit from the outset that whatever our conclusions, nothing can be proven either way? While this game playing, hackneyed, pseudo intellectual enterprise is very much in fashion today, Ortega was already pointing out its development in *The Revolt of the Masses*, in 1930.

> All know that beyond all the just criticism launched against the manifestations of liberalism there remains its unassailable truth, a truth not theoretic, scientific, intellectual, but of an order radically different and more decisive, namely, a truth of destiny. Theoretic truths not only are disputable, but their whole meaning and force lie in their being disrupted, they spring from discussion.[126]

A major component of Ortega's notion of the forms of nihilism enjoyed by mass-man has to do with the negation of ideas precisely because ideas are bound by rules. The "rules of the game," as he

125. Ibid., p. 177.

126. *The Revolt of the Masses*, p. 104.

calls it, bind the thinker to a strict checkmate. Ideas are properly speaking, according to Ortega, the result of a desire for truth. As such ideas come about as the consequence of wanting or desiring knowledge. This is an important contribution on the nature of revolt because it essentially ties ideas to the noble man and the lack thereof to the realm of public opinion. How can we negate, that is, go against reason and values if we can't truly ascertain what either of these mean in their fullness?

A clear exposition of the ominous nihilism that he predicted is not that the vulgar, as he puts it, thinks itself morally excellent and not vulgar. If this were merely the case we would be no further removed from a part of the human condition that has been recognized for thousands of years. The real concern follows, Ortega goes on to say, "that the vulgar proclaims and imposes the rights of vulgarity, or vulgarity as a right."[127] But why the change, we will ask? The answer, in many respects, explains Ortega, is one of simple prudence: "similarly in art and in other aspects of public life. An innate consciousness of its limitation, of its not being qualified to theorize, effectively prevented it doing so."[128]

The explanation as to why this lack of reason exists as a form of nihilism is not purely epistemological in make up, but rather moral.

127. Ibid., p.104. Ortega goes on to add: "The command over public life exercised to-day by the intellectually vulgar is perhaps the factor of the present situation that is most novel, least assimilable to anything in the past. At least in European history up to the present, the vulgar had never believed itself to have "ideas" on things. It had beliefs, traditions, experiences, proverbs, mental habits, but it never imagined itself in possession of theoretical opinions on what things are or ought to be — for example, on politics or literature," p. 70.

128. Ibid., p.104. This section of *The Revolt of the Masses*, "Why the Masses Intervene?" is essential to understanding the second half of the book where Ortega deciphers the social/political conditions that come as a result of revolt. He continues: "But, is this not an advantage? Is it not a sign of immense progress that the masses should have "ideas," that is to say, should be cultured? By no means. The "ideas" of the average man are not genuine ideas, nor is their possession culture. An idea is a putting truth in checkmate. Whoever wishes to have ideas must first prepare himself to desire truth and to accept the rules of the game imposed by it. It is no use speaking of ideas when there is no acceptance of a higher authority to regulate them, a series of standards to which it is possible to appeal in a discussion. These standards as the principles on which culture rests," p.71.

The desire to destroy all standards creates what Ortega refers to as the true form of barbarism: the absence of standards to which appeal can be made. Implicit in this moral dilemma is the view that questions what basis, then, can there be for the future rule of law? But at this juncture we must also make clear that Ortega's solution to the problem of barbarism imposed by the masses is not that they should follow submissively. The safeguarding of the rule of law is not found in blind submission to the positive laws imposed by the state, but to embrace the cues provided by the leadership of the noble man. True individuality is best represented as a vocation that finds its culmination in self-understanding — innately or otherwise — where knowledge of our ability and limitations contribute to the greater good what exists in proportion to our vocation.

The "reason of unreason" is the opposite of life as-having-to-do, the exigencies and contingencies of human reality — instead unreason is guided by the illusion of liberation. But liberation from what, exactly? *The Revolt of the Masses* frustrates some less accommodating readers because its essayistic style guides more than it "theorizes" and its probing offers no fashionable, ready-made formulas. This work is not interested in social-political canvassing, as it takes a sincere philosophical glance at the foundational ingredient of the social-political: man's essence.

The Revolt of the Masses anticipates what is called today "post-modernity" but which is however an old, all-encompassing framework of nihilism. What is so well expressed by Ortega's work is that without seemingly attempting a pretentious analysis of epistemology he manages to demonstrate the bankruptcy of epistemology in what is considered a "post-modern" perspective. Undoubtedly, Ortega offers, much like Camus — to mention just one other contemporary — an objectivist, classical rendition of the plight of man in the cosmos. As such, Ortega founds his arguments using tools that have served man and humanism rather loyally in human reality: common sense, empirical data, and respect for the essential qualities and principles of differentiation. These are all tools that have, with sustained accuracy, delivered man to truth, or truth to man as the case may be. What Ortega points out, then, is that dating back to the nineteenth century these rational/civil tools have been viewed as moral constraints — repressive shackles — to judge by some "post-modern" animated accounts of intellectual history. The negation of these hitherto civilizing tools seems more motivated by liberating forces than it is by the "deficiencies" of the object of its negation.

Today irrational social/political forces squeeze reason in an attempt to remove the alleged source of this cloak of "oppression." But what has changed? The gravitational pull of the Earth has not changed given the stability of the Earth's relative size and mass. H_2O continues to be a compound of oxygen and hydrogen that will embrace the shape of any container, and that will turn to ice when cold and to steam when heated. Diseases of the human body, whether those known to ancient civilizations or present-day ones, continue to make havoc at the molecular level. What, then, has changed? Here the question turns on our notion of personal freedom — "rights," our emotional desire to compress the exigencies of human reality and perhaps strongest of all — a "post-modern" penchant for indiscretion. Human reality has not changed — that is, how it is construed and apprehended independently of irrational emotions.

The revolt cannot help, Ortega explains, but to turn man on himself in the pursuit of unrestrained passions. This, then, is his visionary insight into what revolt means in the twentieth century. How much that passes for "epistemological" thought today can be best understood as precisely its opposite: Calculated, passionate, social/political sidestepping of human reality remains to be seen. What is clear is that such an inquiry cannot be achieved without rational integrity and intellectual sincerity.

Revolt cannot also help but to become destructive because as an all-encompassing form of nihilism it must of necessity destroy the foundation and fabric of rationality. Given the systematic approach and aims of this unprecedented nihilism, its outcome cannot be anything less than what Ortega viewed as barbarism, or as we have already noted, the destruction of all valuation and objective standards.

One of the more neglected aspects of the revolt, as Ortega presents it in *The Revolt of the Masses*, is that this revolt that tries hard to pass itself off as "individualistic" and which aims at establishing more personal "rights" instead owes its allegiance to collective banality. Because the driving force of this revolt is focused on the destruction of values as vitally accessible units of reality, the past, and all kinds of standards, there can remain nothing left to guide us in our pursuits except the whims of strong-arm, gravity-defying cynicism. This cynicism is an integral form of revolt because in demanding ever-expanding personal rights, an elaborate mechanism of rationalization is construed that alleges that "what I demand of human reality is no different than anything that others, today or yesterday, have sought." Currently, the nihilism brought about by revolt takes a multitude of forms, but their unifying strain is that they all seek an alleged greater moral liberation. There can be no denying that this is a strong argument for immorality embracing nihilism — without calling it such.

This form of revolt, because it is precisely the anti-thesis of genuine individuality cannot help but to eventually settle into the anonymity and protection of the politicization of human reality. This is a dominant way in which mass-man can convince itself that "we are all made of the same moral fiber" and that "we all desire the same things from human existence."

The loss of genuine and vital individuality is one of the reasons why *The Revolt of the Masses* can take place in the first place, Ortega argues. What revolt seeks is shelter in familiar places, commonplace ideas on human reality and the soothing disingenuousness found in crowds. These are the rewards of living up to the level of the times, of joining groups for the sake of "belonging," of being taken up by cutting-edge values, and bandwagon causes. Any sincere study of individuality will not fail to reveal that individuals do not think of themselves as abstractions. Whatever niche individuals create for themselves in the world cannot avoid the mirror effect of differentiation. A form of nihilism that negates the self-grounding power of existential reflection is nothing other than an empty shell — for itself — as well as others. No semblance of individuality can exist when the concreteness of vital human existence is systematically negated.

Genuine revolt in both Ortega and Camus showcases an existential concern for the plight of the self that garners self-understanding. For instance, Camus argues: "Metaphysical rebellion is the movement by which man protests against his condition and against the whole of creation. It is metaphysical because it contests the ends of man and of creation."[129]

This notion of rebellion is no other than a stoic attitude toward life, a kind of understanding that asserts "here I am, now what must I do with my life?" What metaphysical rebellion finds when it fashions reflective thought is no other than a conscious "I" who suffers concrete pain and which can focus its vital energy on reflections on itself as well as the world at large.

Revolt in Ortega's work is essentially the same in scope as Camus', but different in focus. The reason for this is that the revolt that is witnessed in mass man is the anti-thesis of Camus' metaphysical rebellion. However, if this were all that can be said of these forms of revolt, then the comparison would very likely fail. Camus' metaphysical rebellion is said to belong to Ortega's noble man because it is the latter, properly speaking, which suffers the lash of cosmic indignation the most. Thus it is the noble man's life that must grapple

129. Albert Camus. *The Rebel: An Essay on Man in Revolt*. Translated by Anthony Bower. New York: Vintage International, 1991, p. 22.

with life as resistance. The noble man experiences this latter category as a future-oriented reservoir of possibilities. This is why the noble man's life is framed by the realization of limitation in human existence. Understanding cosmic limitation allows for the respect of human reality while simultaneously ennobling the pursuit of human freedom.

Ortega aptly demonstrates in the latter half of *The Revolt of the Masses* how the only vehicle left for the mass man to recognize anything vaguely resembling an inner life is that of the state. Ortega spends the first half of the book proving how the rise of mass man comes about and its essential moral/existential bankruptcy. The second half of the book develops the idea that the state ideally should be under the guidance of noble man, but how it is de facto ruled by mass man. Ortega explains this as such:

> The day when a genuine philosophy once more holds sway in Europe — it is the one thing that can save her — that day she will once again realize that man, whether he likes it or no, is a being forced by his nature to seek some higher authority. If he succeeds in finding it of himself, he is a superior man; if not, he is a mass-man and must receive it from his superiors.[130]

And then, just in case that the reader forgets or strays from his major contention — that is a staple of all Ortega's work — he reminds us: "For the mass to claim the right to act of itself is then a rebellion against its own destiny, and because that is what it is doing at present, I speak of the rebellion of the masses."[131]

For another strikingly clear example of this idea we can turn to Wyndham Lewis, an individualist thinker par excellence. In *Wyndham Lewis: A Portrait of the Artist as the Enemy*, Geoffrey Wagner highlights this Ortegan notion. Wagner writes: "Lewis begins his criticism of the contemporary scene from this last point of view, with the complaint that politics today implies a subordination of the

130. *The Revolt of the Masses*, p. 116.

131. Ibid., p. 116.

intellect to practical ends and is thus inimical to the functioning of the true individual."[132]

This is a considerable point to consider in *The Revolt of the Masses* given Ortega's conviction that politics only manages to skim the very surface of human existence. Wagner adds:

> The reason why our age has become so cravenly political, in this sense, is that the true individual, whom Lewis defines as the abstract or quintessence of the group, with a life accordingly more intense that that of the group, has become lazy; as a result, the group or syndicalist ideal thrives. For the true individual must become increasingly energetic in an age like our own, when the body triumphs over intellect, 'sensation' over mind.[133]

Ortega argues convincingly that the failure, that is, the emptying of man's interiority, has a direct correlation with the rise of political posturing. Thus the culmination of mass man's "direct action" is always measured in the realm of politics. It is through politics that the resentment of mass man for human reality becomes objectified, often as we witnessed in the twentieth century, displaying its most brutal and destructive possibilities.

The political process is a mass process, even in its more civil rendition. Politics invites and garners the ability to say a great number of things to the greatest possible number of people. This may sound innocent in itself, however this also robs us of our ability to remain authentic persons — that is — true to ourselves. The political process by definition must also remain a public process that must observe objective rules, customs and procedures that are of necessity anchored in the past. And since Ortega has made clear that barbarism is the destruction of all values and standards, only then can we come to the realization that to lead, leaders must observe some form of standards that serve as the basis of that leadership. Thus, the problem of nihilism and politics best defines itself in its public

132. Wagner, Geoffrey. *Wyndham Lewis: A Portrait of the Artist as the Enemy*. New Haven: Yale University Press, 1957, p. 32.

133. Ibid., p. 33.

embodiment, a form of liberation that demands objectification — at all cost.

Chapter 7. Authenticity and Borrowed Opinions: The Bloated Ship of State

Ortega's study of the modern state begins in the Middle Ages. The state, he tells us, represents the supreme achievement of civilization. But because the organizational complexities that the state demands require the effort of its citizens for its upkeep, Ortega argues that this tension, which is now exacerbated by mass man, only begins to expand. Hence, the modern state, he explains, eventually proves to be, not the benchmark of civilization, but rather its opposite, the rule of mass man. He writes the following:

> He sees it, admires it, knows that "there it is," safeguarding his existence; but he is not conscious of the fact that it is a human creation invented by certain men and upheld by certain virtues and fundamental qualities which the men of yesterday had and which may vanish into air to-morrow.[134]

The major concern with the state, as Ortega defines the problem, is state intervention: "the absorption of all spontaneous historical action, which in the long run sustains, nourishes, and impels human destinies."[135] But here Ortega continues to develop his Vital-

134. *The Revolt of the Masses*, p. 120.

135. Ibid., p. 120.

ism, or what is also essentially a form of Personalism by stressing that the state cannot be allowed to fall into the hands of the masses. In an ironic twist, Ortega does not allow his analysis to stray too far from what the state should protect, which is in fact what it destroys: individual vitality. The state, in its contemporary manifestation possesses what Ortega calls "an anti-vital supremacy" that serves as "the bureaucratization of life." Ortega's brilliant move in this respect is that of noting that while the state is born of the inquietude and insecurity of noble man, the security that the state eventually achieves is the same security that creates and maintains mass man.

Statism for Ortega comes about as the result of direct-action by mass man who is acting for himself. This condition, we ought not to forget, is the same that Ortega asserts when he talks about authenticity (*ensimismamiento*) and inauthenticity (*alteración*) noble and mass man and existential inquietude and barbarism. The vital/existential categories that Ortega promotes as the seed of human reality, unfortunately cannot help but to be eroded by Statism. He tells us: "Can we help feeling that under the rule of the masses the state will endeavor to crush the independence of the individual and the group, and thus definitely spoil the harvest of the future?"[136]

Hence the question "Who Rules in the World?" that Ortega makes the focus of chapter fourteen of *The Revolt of the Masses* is asked in such a new and bold fashion that it shocks sociologists and other materialist's social scientist. Ortega's answer may reveal and even appear to culminate in questions pertaining to political science, but this is precisely what he avoids at all cost. This is also illustrates the uniqueness of this work of philosophy.

Much too accustomed to viewing human reality in quantifiable ways, materialists seem comfortable with explanations that in many regards supplant the cause with the effect. The obstacle that positivists confront when dealing with social/political questions is that in its concern for social usages, collectivity, nation, state it con-

136. Ibid., p. 123.

verts man into an unrecognizable and debased and debased abstraction. The condition is partly due to the debasement of philosophical questions that have become subsumed by social/political ones. It is also indicative of the destruction of philosophy as a signpost of existential anthropology. Ortega addresses many such questions in *The Revolt of the Masses* — many which are simply ignored wholeheartedly by materialist thinkers who have an axe to grind against man's inherent existential personal worth.

We can cite an example of the aforementioned by demonstrating how existential inauthenticity leads to a revolt against the self in the personal realm, as this is equally commiserate with mass mentality and devil-may-care public opinion. A major reason for this has to do with the essence of truth itself. Ortega's concern in this respect has everything to do with the age-old question of the role of truth in human life both, biographically and existentially as human existence. These questions may sound ethereal or even fleeting to readers who come armed with a materialist social/political notion of human reality, but the truth remains that Ortega's work owes much of the force of his argumentation to empirical facts.

But what is meant by the empirical man-made world is precisely that which Ortega cites as owing its origin to reflective, existential concerns. Ortega does not begin his analysis from a seemingly ready-made, pre-existing state. Instead, his concentration has to do with its origin and upkeep. And since there can be no rational talk of the State without its foundational components, man, then his philosophical analysis of this question is all the more warranted.

The notion of truth as a guide of human existence, Ortega argues, literally has everything to do with philosophical reflection and its history. If we take serious stock in the idea that Philos + Sophia is the love of wisdom, then this leads us directly to the possibility of a supra mantra of truth, wisdom, that is, universal and eternal principles that exist in relation to man. How truly difficult can it be to conceive of an entity in the cosmos that we refer to as Homo sapiens — an entity that can come to possess unchanging knowledge of itself, collectively and through its respective individual manifestations? The history of philosophy is precisely this search. The all-encompassing point of this search is to take this understanding and then apply it to the lives of the individual entities that through their inherent ability have come to possess it, or to others to whom it has been passed down. These entities use their knowledge and understanding to create greater cohesion, stability and ease the contingencies of life in wider circles. This is the purpose of wisdom and its search. Ortega reminds us that eventually this all becomes lost in our incessant embracing of the surface structure of human existence: politics.

In some very profound respects *The Revolt of the Masses* is like a work of physics in that physics attempts to explain the macro in terms of its atomic components, which as we can verify today cannot be denied. Much of Ortega's thought accomplishes the same task by teasing the reader with the importance of all-encompassing questions, when in effect its the quantum level, he suggests, that determines the greater picture. For instance, in *Historical Reason* Ortega postulates the idea that history as we know it today needs to come to terms with the understanding that reason, dating back to the ancient Greeks has given western culture the basis of its *raison d'être.* This understanding, this teleological movement of parts within the whole, Ortega argues, has been dealt a debilitating blow by nihilism. He explains: "In a strict sense, the debate today turns on whether truth is manifest in the individual, in one person, or in collective life, in a people."[137]

Ortega carries this essential philosophical problem to its fruition by demonstrating how this is manifested. The culmination of this question, as he views it, has to do with the rift between public opinion (mass man) and the autonomous thinker (noble man). Ortega never argues that the autonomous individual has the right to go against public opinion. On the contrary, the autonomous thinker

137. *Historical Reason*, p. 155.

does not take on the will of the majority for its own sake. However, this question — the nature and manifestation of truth — is best answered through the realization that it is the vocation of autonomous thinkers to engage the solitary pursuit of truth. He explains:

> Here is the second general observation: it does suffice for an opinion to be evident, and hence true, for it to win acceptance. Mankind is not spontaneously open, predisposed, or receptive to evidence. Instead this predisposition is what an intellectual struggles to cultivate in himself: it is his technique and his craft.[138]

We can easily decipher that Ortega's political thought is a classic example of liberal humanism where the individual is the cornerstone of social/political reality, and not the State. Being the genuinely liberal thinker that he is makes Ortega something of an anomaly today, in what has proven to be a post 1968 illiberal western world. However, his historical position is rather unique. Ortega retains the principles of classical liberal democracy and humanism even after the advent of Marxism and Leninism, totalitarian systems that have been embraced by so many other intellectuals. Because his thought begins to flourish during the anti-humanistic, Marxists "experiments" that marked the twentieth century as nothing short of institutionalized murder, he was able to view for himself the effects of totalitarian Statism. But this has also served to place him much ahead of anti-liberal "theorists" who launched the next stage of revolt. As a warning to "post-modern" historical myopia, *The Revolt of the Masses* has no rival.

Part of Ortega's correct assessment of the new man is precisely the latter's ability to shatter all respect and compliance for logic and reason for personal gain. This is the stamp of mass man, Ortega warns throughout this insightful book. But he also predicted and demonstrated how proponents of "the reason of unreason" needed to destroy traditional standards — consciously or otherwise in order to advance unbridled, self-serving motives. We have already

138. Ibid., p. 155.

mentioned how this applies to specialization, aesthetics and moral standards. This, Ortega warns, is a form of historical mutiny — this unprecedented revolt against the self in the guise of liberation. At the heart of this nihilistic assault stands reason and its ability to decipher truth from nihilism and untruth — not mere error — and to condone all forms of unjustifiable institutions. Ortega is quick to grasp that once that this arbitrary mechanism of destruction is set to work, the rest of the revolt or liberation, as the case may be, quickly follows. What seems so instructive in his analysis of mass man is that he paints a picture of sophomoric sensualism, that when read seems too baroque and surreal to be accepted as true. This seems very much like Augustine's notion of time where he argues that he understands time perfectly well when not asked to explain it, but that he finds himself at a loss to explain it otherwise. Ortega did say that Europeans were quickly getting used to living with decadence.

However much that Ortega describes mass man as self-interested, he does not mean this in an individualistic, libertarian way, as we can see. Mass man's grab-bag attitude toward life is the opposite of genuine individualism, as we shall see later. What mass man accomplishes so well and seemingly fruitlessly is to lose himself in the anonymity of the mass, of public opinion. Like a moral chameleon, changing opinion to suit different situations, mass man gives up his autonomy in what it considers a slick, self-survival manner. We are hard pressed to find a greater example of inauthenticity. Ortega reminds us:

> Man lives habitually immersed in his life, a castaway in it, dragged along moment by moment in the rushing torrent of his destiny; that is to say, he lives like a sleep-walker who is only interrupted by brief flashes of clarity when he glimpses the foreign face of the fact of his life, just as lightning, with its sudden glare, allows us to perceive, in the blinking of an eye, the dark center of the black cloud from which it issues.[139]

139. Ibid., p. 192.

All forms of mass appeal, whether institutionalized or existing as transitory specters of reality, resemble a dark cloud under which man goes to seek shelter from himself. The State is perhaps the best example of this. This essential darkness lies at the very center of what Ortega means when he asks, "Who rules in the world?" The immediate answer is not the rulers: those who exercise material power, but rather mass man "by basing his rule essentially on any other thing than public opinion."[140] The State, Ortega adds, is the state of public opinion. Thus public opinion in Ortega's view is like a great machinery that comes about as the result of a life lived with a mass man mentality. In return, the State allows mass man the illusive notion of having opinions. Public opinion is a supreme paradox, according to Ortega, because even though public opinion is no opinion at all because it does not issue from any one mind, it, like mass man at the personal level will not allow for any individual thought to ferment.

Ortega argues that if individuals can live and act as mass man, then naturally entire groups which make up greater social units will equally exert their will. Concerning mass man, Ortega has the following to add: "I called him the mass-man, and I observed that this main characteristic lies in that, feeling himself 'common,' he proclaims the right to be common, and refuses to accept any other superior to himself. It was only natural that if this mentality is predominant in every people, it should be manifest also when we consider the notions as a group. There are then also relatively mass-peoples determined on rebelling against the great creative peoples, the minority of human stocks which have organized history."[141]

"The Whole World — Nations and Individuals — is Demoralized"

"Who Rules in the World?" is a chapter that stretches for sixty-one pages. As the penultimate chapter of the book, a work which

140. Ibid., p. 126.

141. *The Revolt of the Masses*, p. 134.

many not careful readers has labeled a work of social/political philosophy — Ortega reiterates his conviction that the essential questions of the book are nothing less than the metaphysics of moral quality.

In many respects it has escaped many who have commented on this book that Ortega begins this chapter with the clear admonition that the revolt of the masses is a historic occurrence that is the natural culmination of all previous cultural developments and civilizations. *The Revolt of the Masses* is the closing of a circle, Ortega argues, where the beginning and end of moral/existential strife meet, the latter not recognizing its allegiance to the former. And from this wearing down of man's cosmic sense of awe — the vitality necessary to live our day-to-day lives as a form of resistance — what is left instead, Ortega warns, is "the radical demoralization of humanity." This dehumanization takes varied forms — actually, it is all encompassing — but the two supreme examples of this final stage of dehumanization are: the emptying of man's existential reservoir and the politicization of all aspects of human life.

However, these are translucent realities that are confused and assumed by the majority to be aggregates of human life itself. Ortega argues that the aforementioned categories are infrahuman, and as such also anti-humanistic. He writes: "for a time this demoralisation rather amuses people, and even causes a vague illusion. The lower ranks think that a weight has been lifted off them."[142] Thus Ortega's analyses of resentment is striking because he is quick to notice that man's inherent capacity for resentment is always such that it is directed at the metaphysical qualities in human existence. But under such an aegis, who can be the source of vindication? God? Not God any longer, given that that stage of resentment is already exhausted, and God is now dead. The assault on cosmic reality, let us say, must now bear all of the weight of man's resentment. Political "solutions" must now be bargained for. The utopian notion then proclaims that what is "broken" through natural proclamation must now be made

142. Ibid., p. 136.

right with an iron fist. We must make man happy. We must make man happy even if this means his own demise. Ortega continues:

> This is the terrible spiritual situation in which the best youth of the world finds itself today. By dint of feeling itself free, exempt from restrictions, it feels itself empty. An "unemployed" existence is a worse negation of life than death. Because to live means to have something definite to do — a mission to fulfill — and in the measure in which we avoid setting our life to something, we make it empty.[143]

Thus begins Ortega's decipherment of our existentially stunted, juvenile and demoralized age, a sophomoric tide that celebrates its greatest achievement in the sweeping tide of dominant "youth culture." We have already made mention of his notion of "post-modern" man as a spoiled child who makes incessant demands irrespective of a sense of duty — to itself, others or truth itself. In this chapter we witness Ortega's assessment of the unearthing of the natural course of time, when he argues that our empty fascination with the "now" makes the near past seem even older. This is why he states "if we had no children, we should not be old, or should take much longer to get old."[144]

The demoralization that Ortega has in mind is a negation of duty, of personal destiny, of reality or resistance, in short, of our need for personal transcendence. Here Ortega offers the notion that to lead is not equivalent to "how to fill the plate," but rather to instruct as to the importance of "why we need plates in the first place." Leadership, he tells us, is nothing other than the necessary, and hence resented task of keeping man from "wandering aimlessly about in an empty, desolate existence." This is certainly true of the nuclear family. What makes us refute the idea that this is equally true of aesthetics, moral fulfillment or the role of the State?

Of course, one of the most poignant realities that Ortega points out in this work is the need to safeguard our existential/vital inner

143.Ibid., p. 136.

144. Ibid., p.135. A fine and timely example of this moral/spiritual lack of personal realization can be found in the vast array of "cosmetic" surgery that is readily performed today.

life from the objectification of public opinion. But this is a paradoxical task, and probably has always been. Socrates is a fine case in point. While Socrates reflects and feels a certain way toward life, he remains instructive and wise. But when Socrates ventures out into wider circles to communicate his uncovering the principles of human reality — he now begins to elicit the wrath of those for whom he has created a mirror-self. The paradox is universal and seemingly eternal, as far as human existence is concerned: either knowledge and wisdom of human reality do not exist, being mere chimeras, vague illusions, in which case no one knows anything, or it does exist, but is only attained by individuals who make this wisdom the modus operandi of their life. For mass man, Ortega warns, either way is forbidden given that this type of person cannot attain wisdom because of their refusal in the form of: "who are you to tell me about the ways of wisdom?"

This essential theme that serves as the cornerstone of *The Revolt of the Masses*, and which I have already stated, also makes its appearance in his book *The Dehumanization of Art and Other Essays on Art, Culture, and Literature.* He writes the following thus widening the historical depth of *The Revolt of the Masses:*

> A time must come in which society, from politics to art, reorganizes itself into two orders or ranks: the illustrious and the vulgar. That chaotic, shapeless, and undifferentiated state without discipline and social structure in which Europe has lived these hundred and fifty years cannot go on. Behind all contemporary life lurks the provoking and profound injustice of the assumption that men are actually equal. Each move among men so obviously reveals the opposite that each move results in a painful clash.[145]

While this passage is written in the context of art — the nature of the artist, his artistic vision, etc., Ortega nevertheless recognizes that to live is already to live one specific form of existence that cannot avoid its differentiation from everything.

Ortega proves the importance of differentiation in human existence by demonstrating how this is precisely what mass man avoids

145. *The Dehumanization of Art*, p.7.

in its embrace of collective structures. Because differentiation is the basis of existential autonomy and authenticity, Ortega can develop the major characteristics of mass man from mass man's negation of self-imposed standards. Part of this negation has to do with Ortega's view that life is a perpetual series of difficulties that require our attention in the form of reflective thought. Of course, situations, he insists require that we act on them, but action cannot be blind or devoid of some level of reflection. Mass man cannot bring any constructive solutions to a given situation because it finds it imperative to cut corners and thus skip this basic requirement of action. But even most importantly to Ortega's thought is his idea that what truly distinguishes mass man from noble man is practical reflection. The latter comes about through an ongoing, life-long engagement with human reality, not its negation.[146]

Demoralization in Ortega's view is encountered as a cross-national phenomenon that pins man against history, the inventions that have sustained him and the traditionally accepted and presently discarded moral, intellectual and spiritual notions that he no longer accepts. Demoralization is a complimentary condition in the unleashing of mass man's pent up program of moral liberation that it envisions for itself. When he speaks of demoralization, Ortega does not concern himself with this or that form of aesthetics, morality, intellect, or otherwise. His analysis goes beyond any of the current areas where the demoralization may manifest itself. Instead he goes to the metaphysical underpinnings of 'rebellion,' 'revolt,' and moral 'liberation' that can only become manifested at a given historical moment. *The Revolt of the Masses*, we must reiterate, is a work

146. *The Revolt of the Masses*. Ortega writes: "The world at the present day is behaving in a way which is a very model of childishness. In school, when someone gives the word that the master has left the class, the mob of youngsters breaks loose, kicks up its heels, and goes wild. Each of them experiences the delights of escaping the pressure imposed by the master's presence; of throwing off the yoke of rule, of feeling himself the master of his fate. But as, once the plan which directed their occupations and tasks is suspended, the youthful mob has no formal occupation of its own, no task with a meaning, a continuity, and a purpose, it follows that it can only do one thing — stand on its head," p. 133.

of essences, of metaphysical and anthropological qualities and how these inform a cosmic entity that is referred to as Homo sapiens. The greater problem that informs twentieth century mass man is that it has inherited a ready-made world where nothing is autochthonous, a condition that seems rife for nihilism. Ortega explains:

> Human life, by its very nature, has to be dedicated to something, an enterprise glorious or humble, a destiny illustrious or trivial. We are forced with a condition, strange but inexorable, involved in our very existence. On the one hand, to live is something which each one does of himself and for himself. On the other hand, if that life of mine, which only concerns myself, is not directed by me towards something, it will be disjointed, lacking in tension and in 'form.' In these years we are witnessing the gigantic spectacle of innumerable human lives wondering about lost in their own labyrinths, through not having anything to which to give themselves, all imperatives, all commands, are in a state of suspension. The situation might seem to be an ideal one, since every existence is left entirely free to do just as it pleases — to look after itself.[147]

Ortega's idea for moral leadership demands that we come out of ourselves and into the greater world. What this essentially says is that noble man cannot be locked up in itself and lead, given that he recognizes that life cannot be lived in a void. This comes about as the result of creative focus that understands the importance of passing time in human life. Reasons for this have to do with the notion that life is resistance. As such, life engages us to overcome, to transcend while simultaneously employing our time and vital energy in tasks that we can control. This is a substantial moral point that Ortega develops that places the burden of civilization on conscious life: noble man. He argues:

> It is only illusion of rule, and the discipline of responsibility which it entails that can keep western minds in tension. Science, art, technique, and all the rest live on the tonic atmosphere created by the consciousness of authority. If this is lacking, the European will gradually become degraded. Minds will no longer have that radical faith in themselves which impels them, energetic, daring, tenacious, towards the capture of great new ideas

147. Ibid., p. 141.

> in every order of life. The European will inevitably becomes a day-to-day man.[148]

Above all, Ortega is correct in arguing that demoralization does not take place just because someone has referred to it as such. He refutes the academicians' penchant for excessively labeling aspects of reality that cannot be intellectualized. The major reason that Ortega cites as a cause of demoralization is not necessarily due to a decline in the convictions of our art, ideas, our lack of moral engagement with human reality, but rather through a decline in the fullness of our lives. This last category takes place — as does all genuinely lived human reality — in the realm of the vital. What transpires in our ability to communicate anything substantial in our dealings with the social/political, aesthetic, intellectual or moral aspects of our lives is always predated by our engagement with ourselves — here's where man's sense of awe, that proverbial notion of wisdom, cannot be denied its grasp on human existence.

Perception of our lived reality, as Ortega points out is the essential characteristic of a poetic sensibility. This is not equivalent to the perception of life through art, and is certainly anathema to whatever rare accuracy of life that can be found in politics. Thus failures at the social/political level cannot negate their greater allegiance to fundamental failures at the metaphysical/existential level.

Having fully realized this, social/political structures ought to exist as mere correctives to the tribulations and suffering brought about by the life ruled by blinders, whether self-imposed or otherwise. This is the extent of the social /political which paradoxically always possess itself as a "plan for life." The problem here, as Ortega views it, is that the social/political realm is that which by dint of its collective, mass appeal can only be the domain of the cynic. The cynic, as we have already noted, fails in his grasp of inner human reality and thus subsumes this with social/political programs of action for human reality. As such, the cynic negates the vibrant vitality that the noble man recognizes as the basis of existential autonomy.

148. Ibid., p. 144.

Mass man negates all semblance of an inner life out of sheer survival instinct proving that there is "safety in numbers." Ortega explains:

> Life is a petty thing unless it is moved by the indomitable urge to extend its boundaries. Only in proportion as we are desirous of living more do we really live. Obstinately to insist on carrying on within the same familiar horizon betrays weakness and a decline of vital energies. Our horizon is a biological line, a living part of our organism. In times of fullness of life it expands, elastically moving in unison almost with our breathing.[149]

Ironically, the fullness of life can become lost in a state of hyperreality when our expectations of the State have doubled and our existential security vanishes under the illusion of permanency. Efficacy in human existence demands that difficulties become tamed with the understanding that new ones will soon arise. This notion is consistent with Ortega's metaphysical view of man as an entity that is steeped in having-to-do (*que-hacer*). However, the program of politicizing all aspects of human existence threatens — in fact, has succeeded — in unseating our existential inquietude with social/political notions that vanquish our essential qualities under the aegis of a homogenized view of human reality.

Demoralization is a form of moral/existential stagnation. Ask many people from varied walks of life the meaning of anything and very often their answer is technical, even precise. But always the definitive and re-occurring commonality in their answer is that of quantification. "Post-modern" man is adept, and as such is also participant in the cult of quantification, statistics, and pigeonholing of truth: "How does my view of reality relate to my neighbor's?" But ask anyone what is his existential understanding of his role in the great cosmos and at once we are unapologetically deferred to the first category: "I don't know. Whatever the State, or the collective, wants of me." Of course, the scope of this reality is often also more subtle than we can verbalize, making our existential prospects seem even the more daunting:

149. *The Dehumanization of Art*, p. 24.

> For the first time the European, checked in his projects economic, political, and intellectual, by the limits of his own country, feels that these projects — that is to say, his vital possibilities — are out of proportion to the size of the collective body in which he is enclosed.[150]

Demoralization is the logical conclusion of the total politicization of life because, as Ortega points out, the "*res publica*, the *politeia*, which is not made up of men and women, but of citizens," cannot vaguely convey anything constructive as to who is this mass-citizen.[151]

150. *The Revolt of the Masses*, p. 149.

151. Ibid., p. 153.

Chapter 8. Mass Man: The Triumph of the New Man?

The crisis that "post-modernity" has brought about is one that has gone out of its way to embrace abstractions with the misplaced, myopic illusion that only as such can life be conquered. This historical turning point is one that has traded vital life for abstract sociological, ideological principles. The tradeoff imbues man with the idea that personal existence seems less of an existential "shipwreck" if only we can construe it as a collective reality: we purchase the same goods, attend funerals methodically and clinically, and embrace all of the trivial banalities that others present us with. The possibility of stoicism for "post-modern" man seems as misplaced as truth, wisdom and personal loyalty. Man has become a co-conspirator in the mania for abstract sociological "theory" and ideological expediency.

Mass man's disillusion with human reality is created in great part from the cumulative effect of compounding abstraction atop abstraction. Today vague, cynical and sinister abstractions are passed off as truth — conveniently filling in the spaces vacated by personal responsibility and moral courage in lieu of vital existence.

If the world today seems smaller, its borders more readily accessible by television cameras, and if the lure of space exploration has become boring and "routine" — an open book — these, Ortega argues, are all signs of living in a moral/existential suspended animation — the embodiment of a science fiction existence. Ortega is correct in arguing that today reality seems so fragmented, yet so clear to the perspective of mass man because "abstract things are always clear."[152]

Abstractions are clear, impersonal, and easily incorporated into patterns and political programs that can be shared at will. When Ortega argues that it is the noble man who is perplexed by his own existence and who must struggle for understanding, he is making a metaphysical/existential statement on the nature of man. This same line of thought is once again reflected in "Who Rules in the World?" when he offers a diagnosis of abstraction vis-à-vis concrete vital life. Ortega does not allow us to forget that what we initially encounter are not the abstractions posited by science or sociological theories, but rather life — vital, pulsating and perplexing individual human existence.

The "man with the clear head" or, said another way, the temperate person, Ortega tells us, looks life stoically in the face and does not hide in collective structures. In doing so, noble man comes to the realization that life is difficult. The irony here is that the more that this individual launches himself at the mercy and fury of human reality, the more that he comes to view himself as shipwrecked. Yet this existential condition does not defeat noble man, instead it forces him to embrace "tragic, ruthless" reality. This, Ortega emphasizes, is no less than an act of salvation; it is not an exercise in abstraction. This, then, is the philosophical pathos that informs *The Revolt of the Masses.* Section seven of this chapter serves as Ortega's analysis of abstraction and concreteness. The important questions that *The Revolt of the Masses* raises are all in light of the aforementioned comparison.

152. *The Revolt of the Masses*, p. 156.

In addition, Ortega has argued that the State ought not to be an abstraction, given that it is the willing coming together of individuals. Hence the book is a further explanation of what happens to man when he assumes no responsibility for his own existence. "What is really confused, intricate, is the concrete vital reality, always a unique thing," he writes.[153]

This is a rendition of man naked against the background of the cosmos — a universal and timeless theme that has been with us from the time of pre-historic cave painters to the dawn of the atomic age and the computer chip. Abstractions and theories alike have not been able to negate this vital reality, only to dislodge it. Ortega adds:

> The man who is capable of steering a clear course through it, who can perceive under the chaos presented by every vital situation the hidden anatomy of the movement, the man, in a word, who does not lose himself in life, that is the man with the really clear head.[154]

The reason that Ortega's thought seems so irritating to "theorists" and people who promote a pastoral and collective vision of man — regardless of how veiled this vision continues to be — is because Ortega presents man as an individual who through his willed moral courage collects the essences of human existence, as if through a sieve. There is no collective blanket that, once placed over human reality, can give us an exhaustive vision of man — this, instead is the make-believe work of abstraction. This is also another respect in which Ortega can be considered an existentialist thinker.

The same way that Ortega's thought oscillates between the dual tensions of *ensimismamiento* (authenticity) and *alteración* (inauthenticity), we also witness this tension exercised in the personal and public. Ortega leaves no place for man to hide himself existentially under the cloak of the State and still manage to retain his autonomy.

153. Ibid., p. 156.

154. Ibid., p. 156.

To live is to find oneself marooned in an objective reality that attempts to objectify us at every turn in the road.

Hence, "these are the only genuine ideas; the ideas of the shipwrecked. All the rest is rhetoric, posturing, farce. He who does not really feel himself lost, is lost without remission; that is to say, he never finds himself, never comes up against his own reality," he explains.[155] This is consistent with his statement in *What is Metaphysics?* that metaphysics exists for those who need it. What he means by this is that human existence can never become consumed by external reality so that it becomes invisible or translucent to itself.

The importance of differentiation cannot be underestimated in Ortega's work because it is from this recognition that the State is formed. The State, according to Ortega, is not a spontaneous coming together of people. "The state begins when groups naturally divided find themselves obliged to live in common. This obligation is not of brutal force, but implies an impelling purpose, a command task is set before the dispersed groups," Ortega argues.[156] The recognition, then, of differences is the motivating force behind the impetus for the formation of the State. This is an important factor on two counts: first, because it is a recognition of existential qualities, and secondly because the purpose of the State is to promote the continual possibility of the development of individual essences. The problem, as Ortega views it, begins when this initial impetus is quenched and is not re-introduced to succeeding generations.[157]

Ortega's classical notion of the State is thus a depository of man's greatest plans and ambitious. The State is not the result of a top-heavy bureaucratization — this comes later in its development

155. Ibid., p. 157.

156. Ibid., p. 162.

157. Ibid., p.162. This is an essential point in Ortega's understanding of both, a study of mass man as individuals, as well as participants in the State. In this section of the book Ortega spends a great deal of time explaining how the State is not a composite of blood ties, language, or "natural frontiers. The State is instead, "State and plan of existence, programme of human activity or conduct, these are inseparable terms. The different kinds of state arise from the ways in which the promoting group enters into collaboration with the others," p. 10.

— but rather is the space to promote individual, future-oriented goals.

Ortega's equation of nihilism and demoralization is linked to a demoralized Europe. He argues that the greatest components of this demoralization are the politicization of all aspects of human existence and that the "life of the world has become scandalously provisional."[158] Inherent in this nihilism is the condition of mass man's world-weariness, if not life-weariness. Man today, Ortega tells us, is morally and spiritually tired. The individual effort that it takes to lead genuine, sincere lives has been supplanted by a "mere falsification of life."[159] This form of fatigue is symptomatic of an existential drying up of sorts that now finds itself directionless. In addition, Ortega's argument throughout *The Revolt of the Masses* makes the point that "revolt" is rooted in an existential crisis. What is witnessed at the surface of life: social/political life, is always the very last to demonstrate this crisis.

"We Arrive at the Real Question"

The last section of *The Revolt of the Masses* revisits the question of just who is responsible for ruling in the world, especially in Europe, which Ortega specifically cites. The problem of ruling is that until a new standard plan for life is obtained our moral bankruptcy can only promote blind action for its own sake. Thus the end of the book

158. Ibid.,162. He continues: "Everything that to-day is done in public and in private – even in one's inner conscience — is provisional, the only exception being certain portions of certain sciences. He will be a wise man who puts no trust in all that is proclaimed, upheld, essayed, and lauded at the present day. All that will disappear as quickly as it came. All of it, from the mania for physical sports (the mania, not the sports themselves) to political violence; from "new art" to sun-baths at idiotic fashionable watering-places. Nothing of all that has any roots; it is all pure invention, in the bad sense of the word, which makes it equivalent to fickle caprice. It is not a creation based on the solid substratum of life; it is not a genuine impulse or need. In a word, from the point of view of life it is false. We are in presence of the contradiction of a style of living which captivates sincerity and is at the same time a fraud. There is truth only in an existence which feels its acts as irrevocably necessary," p. 181.

159. Ibid., p. 182.

plays close attention to man's existential/moral crisis, while the beginning coupled this with the phenomenon of agglomeration.

He begins this section: "This is the question: Europe has been left without a moral code. It is not that the mass-man has thrown over an antiquated one in exchange for a new one, but that at the center of his scheme of life there is precisely the aspiration to live without conforming to any moral code."[160]

The "real question" all along has been an existential/moral drying up of man's pathos, as we have already pointed out. Ortega's thought does not mince words or does it settle for a vague and comfortable political correctness. *The Revolt of the Masses* is a book that is effective on many levels, but its strongest philosophical contribution is the signaling out or forewarning of the all-encompassing ravages of the making of "post-modern" man. "Post-modern" is a new word for an ancient moral malaise: a cynical nihilism that posits that everything is permitted or that it should be. The nihilism of "post-modernism" a moral Pandora's box that does not confine itself to this or that aspect of human life, but rather which, because of its promise of moral liberation, unseats all previously held notions of human life — this, then, is a beast of prey that has no rival.

The all-encompassing content of this new degenerative, nihilistic pathos is also all-promising. The greatest danger, as Ortega presents it, is that the point has already come to pass when even the word "morality" can no longer make sense. The new morality, he tells us, is precisely the badge of immorality. This is the historical consummation of "liberation," — a settling of the score with life itself. Thus, we can argue that a new age of "post-intelligibility" has begun, but who will recognize this, if part of the moral bankruptcy and operating technique of liberation is to cut off all communication with the past both — distant and recent. Ortega explains: "Hence it would be a piece of ingeniousness to accuse the man of today of his lack of moral code. The accusation would leave him cold, or rather, would

160. Ibid., p. 187.

flatter him. Immorality has become a commonplace, and anybody and everybody boasts of practicing it."[161]

A further forewarning of "post-modernism" is the latter's penchant for self-survival where it must by necessity continue to posit and emphasize an ever-expanding list of abstract "rights" as a detriment to verifiable notions of duties and obligations. But why? Ortega asks. The immediately felt and clear danger of this attitude toward human life is the existential/moral turpitude that fashions destructive illusions in order to assuage the reality that human existence is, or can be existentially taxing, and that there are no certainties. Neither does human existence necessarily present man with moral rewards based on lived virtues. This existential/moral fog may be the initial stage of this "post-modern" condition, but its bastard off-offspring is necessarily the politicization of all aspects of human life.[162]

The politicization of human reality has effectively leveled all semblance of a healthy pathos for human existence, precisely because all modes of life and all moral codes now appear interchangeable and thus also dispensable. This inversion occurs because, as the existential/moral revolt presents itself as the advocate of the people (technically the disenfranchised "mass-man'), it has not only gained

161. Ibid., p. 187.

162. Ibid., p.187. Ortega goes on to explain: "If we leave out of question, as has been done in this essay, all those groups which imply survivals from the past — Christians, Idealists, the old Liberals — there will not be found amongst all the representatives of the actual period, a single group whose attitude to life is not limited to believing that it has all the rights and none of the obligations. It is indifferent whether it disguises itself as reactionary or revolutionary; actively or passively, after one or two twists, its state of mind will consist, decisively in ignoring all obligations, and in feeling itself, without the slightest notion why, possessed of unlimited rights. Whatever be the substance which take possession of such a soul, it will produce the same result, and will change into a pretext for not conforming to any concrete purpose. If it appears as reactionary or anti-liberal it will be in order to affirm that the salvation of the state gives a right to level down all other standards, and to manhandle one's neighbour, above all if one's neighbour is an outstanding personality. But the same happens if it decides to act the revolutionary; the apparent enthusiasm for the manual worker, for the afflicted and for social justice, serves as a mask to facilitate the refusal of all obligations, such as courtesy, truthfulness and, above all, respect or esteem for superior individuals."

the necessary adherents to topple all standards, but it has placed itself in positions — formerly of self-sacrifice and responsibility, now merely of "power" — that will guarantee the expansion of this nihilistic anti-humanistic *geist*.

A fine example of this politicization has been expanded to include the conditions imposed on youth by the laws and principles of life itself. When people refer to themselves as youthful or young, Ortega explains, what they are seeking is the privileged position of having fewer responsibilities and duties than they have rights. Youth, because it is not in a position of genuine accomplishment, has always "lived" on credit." This is an enviable position, according to the new man. Why not make life into a perennial youth? Ortega explains:

> The mass-man is simply without morality, which is always, in essence a sentiment of submission to something. A consciousness of service and obligation. But perhaps it is a mistake to say "simply." For it is not merely a question of this type of creature doing without morality. No, we must not make his task too easy. Morality cannot be eliminated without more ado. What, by a word lacking even in grammar, is called amorality is a thing that does not exist. If you are unwilling to submit to any norm, you have, *nolens volens*, to submit to the norm of denying all morality, and this is not amoral, but immoral.[163]

By recognizing that neutral amorality cannot exist, Ortega essentially forces the denizens of "post-modern" nihilism to accept their notion of immorality as arbitrary forms of distorted "morality." Thus ends *The Revolt of the Masses*:

> It is a negative morality that preserves the empty form of the other. How has it been possible to believe in the amorality of life? Doubtless, because all modern culture and civilization tend to that conviction. Europe is now reaping the painful results of her spiritual conduct. She has adopted blindly a culture which is magnificent, but has no roots.[164]

163. Ibid., p. 189.

164. Ibid., p. 189.

Chapter 9. The New Man: Parody of Genuine Individualism

In 1930 Ortega found himself in the peculiar situation of trying to answer questions of daily life both of an empirical nature and those that pertain to existential inquietude, with the tools supplied to him by philosophy. But this is hardly strange, or even innovative, some readers may suggest. Ortega's age was perhaps the last genuine age of free thinkers and intellectuals — people who addressed the eternal questions of man without allegiance to the special interest of ideology: the role of human existence vis-à-vis the cosmos, mortality, the qualitative differences that mark man's existential/moral perspectives and the exercise of man's metaphysical, primal freedom and how these affect man's behavior amidst others. The distinctive quality of Ortega's generation, dating up to the late 1950s, is the ability and desire to engage philosophical questions with the necessary rigor and intellectual integrity that these eternal questions require. Ideology, as Ortega argues throughout his work, is the activity of morally short-sighted, one-track mind pundits. The central questions of human existence, as he tells us, whether addressing the nature of purpose and cultural valuation, technology or aesthetics, may in effect impact the social-political arena, however, they must never be forged there.

Hence, Ortega's major difficulty in writing *The Revolt of the Masses* was how to keep a broad perspective of human existence and at the same time pay allegiance to the overabundance of empirical knowledge that was available to him. This question, then, is hardly one of knowledge, but rather of understanding. This is precisely his major contention in his critique of specialization, for instance. Ortega brings to *The Revolt of the Masses* his understanding of neo-Kantian metaphysics, phenomenology and positivism. But most importantly his early thought on Perspectivism and later on existentialism allow him to showcase a mode of philosophy of life that attempts to retain the essential inner complexity of human beings without sacrificing vitality to abstraction. This negation of abstraction is perhaps the single strongest motivation and indicator of Ortega's philosophical enterprise. In this respect, we can view Ortega as following in the philosophy of life movement that can be traced to Dilthey, Simmel and Unamuno. Thus, it becomes easy to realize why he views the social/political arena and the politicization of life as antithetical to the safeguarding of vital life.

A Dearth of Sincere Sentiment

Ortega has referred to metaphysics as the construction of the human world. Starting with our circumstances, whether internal or external, we must construct a vision of life. But this is achieved first and foremost from a desire to salvage our circumstance, as he explains. The result is no less than the creation of personal existence. This is the meaning of his notion of the existential category, having-to-do (*que-hacer*): "Man, every man, must at every moment be deciding for the next moment what he is going to do, what he is going to be."[165]

165. *Man and Crisis.* "This decision only he can make; it is not transferable; no one can substitute for me in the task of deciding for myself, in deciding on my life. When I put myself into another's hands, it is I who have decided and who

Justification of personal existence is a staple ingredient of existential philosophy. Sartre argues for radical freedom — a form of freedom that must assert itself at all cost — while Camus defends metaphysical revolt, and Marcel the sincere daily renewal of religious belief. All of these forms of existential thought have their own level of complexity and difficulty.

Justification for Ortega cannot be viewed as blind decision-making where we decide on a whim or out of a "career" move what our life (or we) ought to be at any arbitrary moment. Instead this form of existential decision presupposes, whether consciously or otherwise, the anticipation of our future. As such, this is a question that presupposes that we understand and are aware of our vocation, whenever this is the case, and pose to ourselves the problem of our own individual existence, as Ortega explains. But existential concerns quickly demonstrate their central position in human existence by forcing us to take stock of just what we mean by notions such as: "the world," "man," "society," etc., at every step of our reflective questioning. The web of existence in which we find ourselves widens in direct proportion to our questions.

Thus, to live, we must be armored with the understanding that to do so is to "interpret life," or at least to become armed to do so. This is the kind of entity that man is. Ortega understands that this is a fair description of man based on the law of averages. He anticipates, in fact points out, that reflection on existential questions is neither possible for most people nor is it necessarily a pleasant activity. The question then seems to be, what ought one to make of those who refuse or who have no aptitude to grapple with questions of an existential nature? Rather than tackling this question as if everyone is a prototypical first man — which in many respects we might be correct to assume — Ortega fashions this question in terms of societal structures placed on man. While he argues that everyone ought to busy himself with an interpretation of life, this is not what any

go on deciding that he will direct me; thus I do not transfer the decision itself, but merely its mechanism," p. 23.

modicum of empirical data demonstrates. This is lamentable, especially given the revolt of the masses, because this activity is not a luxury, elective form of cosmetics but rather a primary condition of man's being. The problem of life-as-preoccupation is not that preoccupation is entirely absent in lives that refuse to engage existential reflection, but rather that preoccupation is diverted into other mundane and often banal channels.

Of course, in an age that negates all forms of hierarchical structures of reality, the primacy of existential authenticity is also denied. The critics will scoff that one man's authenticity is another's depravity — this is just a matter of self-preservation in a nihilistic age. However, given that Ortega cites the importance of existential proximity, of sentiment to oneself, and not the external world, it then becomes arbitrary to give credence to any valuation that does not differentiate between qualitative degrees. The importance of existential reflection in *The Revolt of the Masses* can now be seen to serve a central role in Ortega's analysis of mass societal structures. Ortega demonstrates that lack of conviction is never a total absence, but rather only the employment of selective convictions.

> For when each one of us asks himself what he is going to be, and therefore what his life is going to be, he has no choice but to face the problem of man's being, of what it is that man in general can be and what is it that he must be. But this, in turn, obliges us to fashion for ourselves an idea, to find out somehow what this environment is, what these surroundings are, this world in which we live. The things about us do not of themselves tell us what they are. We must discover that for ourselves.[166]

What is important to Ortega's notion of existential aridness in *The Revolt of the Masses* cannot be measured with sociological statistics — for, after all, the raw material or data collected by sociologists is not only how people act in social structures — but rather also in the inescapable existential data that serves as the stuff of society. The main reason that Ortega does not build his arguments around social/political structures has to do with the time-proven

166. Ibid., p. 23.

futility of such a narrow enterprise. Regardless, Ortega has a high regard for empirical evidence, a rather interesting irony in itself, given that most of the materialists and positivists whom he refutes ignore the stranglehold of politicization and ideology on their respective thought.

Also part of this irony is the fact that Ortega begins the book by citing historical examples of the masses and their ascension to power. The book begins with the question of agglomeration or what he refers to as "plenitude." He also grapples with the question of plenitude in terms of technology, or what is called applied science today, but which Ortega called technicism. And then there is also an exploration of the nature of the State as a form of mass-directed phenomenon. But always embedded in and guiding these external aspects of the objective world, he tells us, is the unadulterated essence of human beings, an entity that is much more preoccupied with essential metaphysical structures than the zoological moniker "Homo sapiens" may indicate. The distinguishing feature of man as a cosmic entity is not how man becomes adopted to his environment, but rather how this existential being comes to know itself in its total circumstance. The former portion of this question is a dead end, for today, we know full well how man can adopt to technology, moral depravity, gulags, cancer and totalitarian social/political conditions, for instance. In point of fact, we can verify that some people are happy to behave like Pavlov's dog while others perish as martyrs defending truth and personal dignity. It seems that Ortega's existential approach to questions of human essence, moral freedom and autonomy is a more sincere manner of reflection — and certainly less dangerous — than continuing to spin pointlessly in place like a child's top; pseudo answers to ill-reasoned, mere social/political craftiness. Ortega takes philosophical reflection seriously — as being pregnant with possibilities and answers to the exigencies of the human condition. Within the structure of all societies, he clarifies, at the beginning of the book is the question of essence: society and all human interaction is ruled by the motivating fuel that determines mass-man or noble mentality.

The responsibility of the Ortega scholar or his good willed readers lies in their ability to understand the importance of individual autonomy in *The Revolt of the Masses.* While no book is exhaustive, readers must nevertheless engage the author on the latter's terms, otherwise why bother with reading or studying in the first place? We can dismiss "deconstructive" techniques as disingenuous, temporary make-work that act as "symptomatic reading" (as Ortega argues) and as the chief characteristic of "intellectual" mass man in their refusal to engage human reality.[167] Hence, *The Revolt of the Masses* is just one of Ortega's treatises that explores the question, what is man? He does this in relation to both, his own vocation and circumstances, as well as to society. The reason that *The Revolt of the Masses* is his best-known work is in some respects, due to the misguided belief that he solely speaks of social-political concerns.

Consider that in *Man and Crisis* (*En Torno a Galileo*) Ortega offers a much more detailed philosophical argumentation for his social/political ideas. The chapter titled "The Structure of Life, The Substance of History" takes up the question of writing history as a record-keeping enterprise, not of objective, historical events, but of essences: "The fundamental question of history comes down to this, then: What changes have there been in the vital life structure? How and why does life change?"[168]

History, then, is best explained as man's existential drama in finding himself submerged in circumstances that he must attempt to understand, as Ortega suggests, to remain afloat, "to save our circumstance." But circumstances are mute. They do not force us into accepting any given specific course of action. Instead, how we decide to proceed will say something about the degree of engagement that we seek regarding our being. Always having to interpret

167. Tony Judt. "Review of Louis Althusser, The Paris Strangler." *The New Republic.* March 7, 1994, p. 33.

168. Man and Crisis, p. 29.

life — on our own — man makes his own destiny to the degree that he recognizes the possibilities, which is also to say the limitations, of the life that he has been given. Man as radical reality is not as explicit a component in *The Revolt of the Masses* as it is, say, in *Meditation on Quixote*, but without a doubt, noble man comes to this conclusion while mass man is frightened by it.

Human history serves as a common depository of the inexact art of the being of man, a strange entity according to Ortega, to be sure, and one who must make sense of his life in the surrounding world. In a sense, history is the history of everyman, as this is focused under a magnifying lens by the subject. This is an example of Ortega's early perspectivist period where truth is defined as objective, even though effectively viewed as kaleidoscopic — not relative — by all who seek it. This is a prevalent theme in *The Revolt of the Masses* because Ortega argues that what mass man seeks is the convenience of an all too relative rendition of truth, if any, that can be readily accommodated to all manner of personal demands. Of course, it ought to go without saying that the relativist will find it efficacious to take a step further and assume out of survival and self-serving reasons that truth is a "bourgeois" fabrication.

Absent from this rendition of life-as-I-want-it-to-be is a view toward the acceptance of human existence as resistance. *The Revolt of the Masses* is essentially a revolt of the structure of life as we know it. This is why Ortega points out that metaphysics is the science of constructing personal existence — but only for those who need it. This is also why in *Man and Crisis* he attempts to demonstrate that there can be no science of man, given man's fool proof understanding or what is man's limited capacity to fully know his origin. This point is important to *The Revolt of the Masses* on two counts: First, if man were truly omniscience — a notion which Ortega refutes — then this would mean that we are no different than any other closed-ended entity such as a stone. This being true, whatever we needed to know concerning this entity would be static and thus would effectively be known a long time ago. This would also suggest that man is an undifferentiated entity. Secondly, man, Ortega tells us, is the type of entity that finds itself lost both, individually as well as collectively. This reality propels us to understand ourselves, our motives for living and our grappling with personal death. Thus, far from being some sociological country club existence, human life must come face to face with itself, a task that determines the tonic of our destiny, for everyone of our actions negates a series of others that knowingly or not we must discriminate in making choices.

Existential Freedom is a Tenuous Thing

Gabriel Marcel cites Pierre Bost's short story "Monsieur l'amiral va bientôt mourir," where the latter author writes: "If freedom were easy, everything would fall to pieces at once."[169] Marcel's contention is simply that all abstract talk of values and freedom is nothing less than a baseless use of "techniques of vilification" where the latter is used as false adulation of a greater abstraction, "the people."

Similarly, Ortega warns us that human existence is not transferable; life is a frightening task for some people. Now, to refer to life as frightening will no doubt ring Medieval bells in the ears of people who have waived their primal freedom in exchange for collective and progressive values. However, it goes without saying that if we are to pursue intellectual integrity we quickly arrive at the conclusion that man's existential fears cannot be assuaged by social/political posturing. It must be continuously reiterated that *The Revolt of the Masses* is not a work of what Helmut Schoeck refers to "applied sociology" — an ideological make-work discipline that begs the question in its incessant inductive "duplication of effort" — but rather a "pure sociology" of knowledge that takes its cue from ontological differentiation.[170]

169. Gabriel Marcel. *The Philosophy of Existentialism.* Translated by Manya Harari. New York: Carol Publishing Group, 1995, p. 86.

170. See: *Essay On Individuality.* Edited by Felix Morley. Indianapolis: Liberty Fund, 1977. "It seems that sociologists have given up more and more what Lester F. Ward called pure sociology and have imbued their work with what he called

Ortega warns of an ensuing day when the power of the State will easily usurp man's freedom by creating greater expanding collective structures where man can drown his vacuous modern sense of self and existential angst. Thus, revolt is not a revolt against State oppression necessarily, for instance, but rather a revolt against the self and its tired, weary and worn out sensibility that asserts the likes of — "if this is what it means to be a free individual, then I would rather take my chances becoming the Other." In a world where the notion of the "we" has greater primacy than the felt, lived subjective-I, then clearly what matters most is what external forces dictate.

The major reason that mass man, as Ortega tells us, must negate all possibility of individual autonomy in others has to do with the cynical inductive fallacy that asserts that every time I meet another person they are always like me. This makes the characteristic quality of nay-saying the central rallying force of the mass man. This condition resembles a sophomoric vulgarization of preliterate man where it is precisely the self and the fears that it must contend with and that it first encounters. If man is de-humanized and anesthetized to his felt and lived human condition, this loss of autonomy has not come of its own accord, a genuine question of study for pure sociology, which its ideological distant cousin completely negates.

Ortega argues that nineteenth-century civilization is responsible for the formation of mass man. That century is seen as crucial because it is the century that gives birth to liberal democracy and technicism (applied science). Because the age of technicism is also that of specialization, Ortega cites this as the beginning of an age of

applied sociology. As Ward defined it, "pure sociology is simply a scientific inquiry into the actual condition of society." We have now reached a point where pure sociologists, e.g., Kingsley Davis, are attacked for the impediment of (presumably) applied sociology through their work in pure sociology. (Cf: the controversy between K. Davis and M. Tumin, America Sociological Review, August 1953). Many American sociologists today behave as if Ward's definition of applied sociology would have to cover all of their work. They are no longer aware of the difference between epistemological and ontological equality," p. 148.

blinders where the mass man gives up any view of universal principles for the security of focusing on a fixed task.

Human freedom is readily (or potentially) rendered impotent by technicism through mass man's inability to continue to link it with imagination. Technicism robs human freedom of its vital nature, that is, the understanding that freedom comes as a result of consciousness, and consciousness is, at worst, a mechanism of choosing — an inherent system of elaborating reality as hierarchical. Technicism can be defined as taking the seeds of reason (Ortega says pure reason), and leveling it to its most practical and mundane denominator. Now, there is hardly anything wrong with the fruits of science, as we have already noted. What Ortega cites as the major problem of applied science is that it splinters scientific questions into ever-wider areas of specialization. With this comes a loss of coherence for reflection on what Ortega calls a unification of the sciences. Thus, the specialist too, with his blinders and myopic outlook, represents a form of the "intellectual" mass man. Ortega explains:

> By specializing him, civilization has made him hermetic and self-satisfied within his limitations; but this very inner feeling of dominance and worth will induce him to wish to predominate outside his speciality. The result is that even in this case representing a maximum of qualification in man — specialization — and therefore the thing most opposed to the mass-man, the result is that he will behave in almost all spheres of life as does the unqualified, the mass man.[171]

If specialization makes a gestalt interpretation of the cosmos all but an impossible fantasy, it also makes human freedom a superfluous luxury that few will venture to exercise. Specialization in science is necessary because of the demands placed on the scientist by the ever-greater complexity exhibited by the problems that science attempts to solve. But what is often misunderstood or ignored is the effect that narrow perspectives — analysis that is not rounded out with synthesis — can have on human freedom.

171. *The Revolt of the Masses*, p. 112.

Ortega's notion of revolt, as we have previously said, sets the stage for a vacuous revolt in the social/political arena. This is essentially the importance of mass and noble man. An "unemployed" existence, as Ortega has referred to it, is an inauthentic existence because it usurps the Other's freedom without respecting it.

Hence, specialization in one area, say, theoretical physics, is quickly transferred over into spiritual, moral and existential areas of human life. A fine example of this is Einstein's theory of relativity, where what was meant as a mathematical formula to attempt to measure the speed of light in a temporal dimension is consequently taken as a fact in the macrocosm of human values.

But perhaps even more important is the manner in which human existence in the first half of the twentieth century — as Ortega knew it — continues to decentralize or de-universalize man's interpretation of the human condition. There is great irony in this. Freedom for Ortega means the encounter with cosmic resistance, with realities that reason; truth and logic present us with. These are realities — human mortality being a universal example — that beckon for reflection, if not for answers. Thus, existential freedom can attack such questions in at least two ways: First, it must not deny the possibility of encountering unflattering truths. Secondly, it must face the possibility that the rewards of such a search are already contained in the search by allowing for the development of genuine values and thus "saving our circumstances," with the authentic embracing of resistance.

Now, our ability to save our circumstances cannot be realized in isolation: the pursuit of a "career," social standing, the forging of reputation, etc. These particulars, once the thirst for them is quenched, do not showcase any general degree of transcendence. Once met, these goals merely continue to exert greater particular pressures on us that further alienate us from the pursuit of broader understanding. This, Ortega explains, is not existential freedom but rather its negation. By depositing ourselves at the mercy of epistemological and moral scams that attempt to short circuit the resistance placed on us by human reality we instead subjugate the exercise of freedom to my individual desires.

Human freedom, as Camus and Marcel have commented, is not a theory amongst other transitory, fashionable theories, but a vital reality that shows its self-evident fervor the more apt that we become at negating it. When Ortega argues for human freedom he does so from a perspective that equates this with vital life. The opposite of vital life is evidenced in forms of life that prove to be purely artificial fabrications. Therefore, he writes:

> Reason is merely a form and function of life. Culture is a biological instrument and nothing more. When it is set up in opposition to life it represents a rebellion of the part against the whole. It must be reduced to its proper rank and duty.[172]

Pure Reason and Vital Life: A Case of Socratic Irony

Part of the reason that revolt involves giving up our freedom has to do with the embracing of diffuse, abstract principles that can be molded to our every whim. Wherever this mechanism is set in place, we also see a dearth of, if not the destruction of, disciplines and attitudes that formally were aides in grasping the essence of human reality. However, the problem as Ortega views it is rather tenuous, for some seemingly unorthodox reasons. He views the late nineteenth century as the first period after the Enlightenment that reaps the fruits of reason, the ability to adopt to the contingencies of technicism, mechanization, and industrialization. Ortega argues that the Industrial Revolution comes about as the result of pure reason. The period that came after the Industrial Revolution finds itself inundated by applied science in what Ortega argues are very positive ways. However, the major philosophical problem has to do with the existential confusion that material progress can create. He explains:

172. *The Modern Theme*, p. 58.

> To what extremes can this process go? Can reason be self-supporting? Can it get rid of all the rest of life, the irrational, and go on living by itself? To this question no answer could be given at the time: the great attempt had first to be made. The shores of reason had been thoroughly explored, but its extent and content were as yet unknown.[173]

The question: Can reason be self-supporting? only sounds odd because what Ortega means is more akin to: What is the relationship of pure reason to vital life? Pure reason is self-supporting as long as it is a self-contained aspect of intelligence that is ultimately at the service of life. The failure to fully understand this is another of the central characteristics of revolt. Not only does mass man expect pure reason to deliver indefinite and unlimited goods, Ortega argues, but when we give up existential freedom to this unfounded belief, the spontaneous vital quality of human life also vanishes.

The aforementioned condition is no less than the exercise of genuine Socratic irony, he goes on to say. The irony lies in supplanting vitality, spontaneity, and common sense, first hand empirical observation, which he refers to as a primary movement, for a secondary one that is dictated by pure reason. If reliance on empirical observation of the world around us is an essential condition that leads to understanding first principles, then paradoxically this practical engagement ought to be reinforced, not abandoned. The tradeoff for mass man, Ortega argues, is hardly a moral dilemma: if human existence exhibits great resistance to my every whim, and applied science seemingly throws a blanket of protection over all aspects of my life, then why not abandon my existential concerns to the former? Mass man resents all forms of hierarchical authority, including that which will assuage his worries. Ortega argues:

> Socratism, or rationalism, begets, on identical grounds, a double life in which our non-spontaneous character, or pure reason, is substituted for our true character, or spontaneity. It is in this sense that Socratic irony is used. For there is irony in every act by

173. Ibid., p. 56.

> which we supplant a primary movement by a secondary, and instead of saying what we think, pretend to think what we say.[174]

Ortega's major challenge to man in the twentieth century is to pose the question of how to embrace life while simultaneously enjoying the fruits of science. This question is one that he explores in his book *The Modern Theme*. Hence twentieth-century man is defined in part by an understanding of the uses and limits of pure reason and its offshoot, applied science, without negating or repudiating it. What is the terminus of science? Ortega argues, is now the most important question. The modern theme, then, "comprises the subjection of reason to vitality, its localization within the biological scheme."[175] This new form of culture is what Ortega calls vital reason, or the embracing of life by reason. Whether mass man can attain to such enlightenment remains to be seen given the delicate balancing act that Ortega suggests must take place between life and rationalism.

Irrationalism, Sensualism and the Triumph of Despotic Political Correctness

The Revolt of the Masses strikes a chord with conscientious thinkers today because of its sheer breadth of scope and ability to underscore the importance of metaphysics to the human condition. Stretching from ancient history to the 1930s, this chronicles the innermost essence of the problem of agglomeration, the birth of civil society, the expansion of scientific technique and the rise of collective social structures that set aside the inherent worth of individuals.

Ortega's now classic *The Revolt of the Masses* continues to be relevant today; it transcends the immediacy of the problems that afflicted Ortega's Spain. This is why the work has such merit for thinkers interested in questions of culture and the *dignitas* of individuals, for instance.

174. *The Modern Theme*, p. 56.

175. Ibid., p. 58.

Ortega took careful stock of the empirical world around him and from there managed to identify and trace the essential traits of human beings as individuals as well as in our societal reality, and he pointed out that human reality in all of its manifestations is brought about by the degree of differentiation found in vital, self-conscious reality. It very often goes unnoticed that Ortega views man as a transcendent, cosmic being and not as a mere socially/politically "determined" being. But we ought to remain cautious of what is meant by the words "transcendence" and "cosmic entity" in reference to his philosophy. Ortega's work demonstrates almost no sign of any preoccupation with God or religion. By transcendence, however, Ortega signifies a realm of excellence, whether moral, intellectual, cultural or spiritual — by which man — noble man — guides his life. What is significant about transcendence is not that it should serve as a guide for life, but that it is a way of ascendance toward one's vocation. Implicit in this, of course, is the recognition of one's vocation. This is a theme that Ortega does not allow to fester in a debased pseudo subjectivism, but which he posits instead as a universal principle of human existence. Ortega demonstrates vividly how the universal is contained within the individual and how this entity is responsible for discovering its own essence.

By the same token, man, Ortega goes on to say, is a cosmic phenomenon, a being that by its very definition cannot be contained in arbitrary racial, social/political or class structure either by turning its back on its primal freedom or through ideological expediency. It is hardly difficult to conceive of man as a cosmic, self-conscious entity that lives on "borrowed" time in the sphere of material being. All we have to do is imagine any number of people on planet Earth — six billion at the moment — each with a level of differentiation that is particular to its own ability and desire to view itself in this, its cosmic, raw condition. Whether we decide to be carpenters, thieves, writers or wayfarers cannot easily by ascribed to some conveniently prescribed social/political reality. In the end we live much as we die, within ourselves and the view that we have of our existence.

Ortega's existentialist approach in *The Revolt of the Masses* suggests that our view of ourselves has a great deal to do with our understanding of social/political reality. However, as soon as man starts bidding for change on the basis of ideological reasons, he cannot always foretell the effects his ideas will have on others. Revolt occurs when mass man forges ahead with the earlier view of reality in mind.

The anti-vitalist warning of *The Revolt of the Masses* reverberates with forewarning of the anti-humanist revolt that would take place thirty years later, in the 1960s. The irony of the revolt that Ortega simultaneous describes and predicts (in its continual metamorphoses) is that while the revolt is about greater moral permissiveness, it also demands a greater politicization of human life. This seems to be an indication that what revolt really means is simply a manner of institutionalizing "my particular" form of degeneracy. The only way that the greatest common denominators of human existence can be sanctioned is by promising everyone the indefinite expansion of his moral "freedom." This may be a disingenuous and sleight-of-hand manner of further liberating ourselves from the constraints of moral imperatives and even from objective reality, but what is created in the short term is a form of totalitarian, politicized view of human reality that is anything but spontaneous and vital. And since all the forms of vulgarity that beckon for a piece of this historic moral disintegration and which now see themselves as vindicated of the evils of the past — real or alleged — cannot help but to continue to sanction further liberation out of a self-serving motivation. As a form of resentment, the alleged liberation of revolt could not be more fitting a twentieth century staple. The new despotism hails itself as a timely model of universal human suffrage out of the kind of arrogance that only despotism can supply. Robert Nisbet enlightens us in this respect when he writes in *Twilight of Authority*:

> We have seen, alas, the appearance of *ressentiment* that Tocqueville and Nietzsche, among others after Burke, predicted: the sense of the greater worthiness of the law, the common, and the debased over what is exceptional, distinctive, and rare, and, going hand in hand with this view, the profound sense of guilt — inscribed in the works of the new Egualitarians — at the sight of the latter.[176]

Ortega points out is that in order for the lowliness that revolt calls for to become institutionalized one must first take control of the mechanisms that dispense culture. It is essential to first politicize all of human existence and the institutions that this has created in order to discredit tradition and thus demonstrate how the "new" values of liberation are equal if not superior to traditional values. Nisbet adds:

> Hierarchy in some degree is, as I say, an ineradicable element of the social bond, and, with all respect for equality before the law — which is, of course, utterly vital to free society — it is important that rank, class, and estate in all spheres become once again honored rather than, as is now the case, despised or feared by intellectuals. Certainly, no philosophy of pluralism is conceivable without hierarchy — as open as is humanly possible for it to be but not, in Burke's word, indifferently open.[177]

The idea of hierarchy informs *The Revolt of the Masses* as a characteristic of ontological differentiation and not as an arbitrary social condition, as its destruction has proven to be. Ortega addresses the question of hierarchy much as Plato does by citing ontological concerns like: being-becoming, finite-infinite, life-death, and the degrees and qualitative differences exhibited by human reality. Human existence is teleological whether we are cognizant of this. But the life that is self-aware of setting goals and purposes for itself is also an authentic life, a form of existence that is proactive.

One of the strongest attributes of the noble man, as Ortega presents this type of individual, is his ability to always remain on the vanguard of a reflective perspective of his future. Thus, to live is to

176. Nisbet, Robert. *Twilight of Authority*, p. 218.

177. Ibid., p. 218.

direct our lives toward an objective resolution of our spatial-temporal lives. But to live with a view to the future entails — of necessity — a series of convictions and beliefs that express our goals and the vehicle that we utilize in enacting them. This alone is already a condition that removes this reflective person from the run of the mill, easy "thoughts" of the herd. We can say that as far as noble man is concerned, convictions, both heart-felt and vital, are his saving grace. This is another important point to consider in *The Revolt of the Masses* because convictions properly are not forged by the whim of any collective standards, even though they can be passed down. What this means to the life of noble man has everything to do with Ortega's contention that such a life sees existence as a labyrinth that does not easily give away its secrets. Eventually the noble man can come to share his findings with others much like himself but this coming together must take place out of fullness and a surfeit of emotions and not out of emptiness. What the noble man discovers in his search for convictions and genuine beliefs is that human reality conforms itself in a more objective and universal manner than mass man cares to know. To understand the terms set by the human condition is to progress into the real possibility of our transcending them. Ortega argues:

> To have an idea means believing one is in possession of the reasons for having it, and consequently means believing that there is such a thing as reason, a world of intelligible truths, to have ideas, to form opinions, is identical with appealing to such an authority, submitting oneself to it, accepting its code and its decisions, and therefore believing that the highest form of intercommunion is the dialogue in which the reasons for our ideas are discussed.[178]

What did Ortega already see in 1930s Europe to warrant the writing of his prophetic treatise on man in revolt? While the word "revolt" has been used in a large and often vague array of senses, the meaning that the word retains in his classic book today is its original meaning. It is perhaps also safe to say that while revolt of one kind or other has formed part of human consciousness, this nihilis-

178. *The Revolt of the Masses*, p. 73.

tic revolt is such a centralized and fundamental kind that it threatens the very essence of human reality. Revolt for Ortega signifies revolt against the self in a self-serving refusal to come to terms with the human condition. While we can argue that revolt threatens human reality as we know it, it nevertheless cannot help but leave the human condition intact, for the latter is governed by principles that we have no means of changing. We can reject, refuse and refute the human condition, but we cannot change it as long as we remain the entity known as Homo sapiens.

Revolt is said to be a form against the self because it is intent on a wholesale destruction of culture regardless of our being cognizant of this phenomenon. The "direct action" of the masses, Ortega argues, is a destructive and common way in which authority is suppressed. He calls this the "Magna Charta of barbarisms."[179] Revolt is a form of revolt against the self in its ability to destroy civilization, which as Ortega demonstrates, has always been an instructive influence in leveling the use of force. The force used by "barbarians" Ortega argues has always been the *ultima ratio*, or what happens when reason becomes exasperated. "All our communal life," he writes "is coming under the regime in which appeal to 'indirect' authority is suppressed, in social relations 'good manners' no longer hold sway. Literature as 'direct action' appears in the form of insult. The restrictions of sexual relations are reduced."[180] And then, rather than merely citing the excesses of vulgarity he tells us why these forces are so destructive:

> Restrictions, standards, courtesy, indirect methods, justice, reason! Why were all these invented, why all these complications created? They are all summed up in the word civilization, which, through the underlying notion of *civis*, the citizen, reveals its real origin.[181]

179. Ibid., p. 75.

180. Ibid., p. 75.

181. Ibid., p. 75.

Thus, *The Revolt of the Masses* debunks the proponents of total liberation and their claim to a "higher" consciousness and an expanded sense of "social justice" by demonstrating the extent of the contradictions that such self-serving liberation pretends to achieve.

Chapter 10. Mass Man's Existential Revolt and the Future of Human Freedom

The first decade of the twentieth first century finds man a tired, weary entity who believes in nothing, trusts not even himself and possesses no sincerely felt convictions. At least none that are vitally lived and consistently measurable. His behavior gives testimony to a barren interior — that is one that is not a genuine interiority at all — only a hollow, thespian self. This is the culmination of an entity whose viscera have been spilled over the terrain of an abject, disingenuous material condition. This is an entity who has traded its existential freedom for the easy and carefree levity of determinism and all manner of transitory and social/politically expedient "theories." This is the picture of a gullied, hologram man, a carapace, a farce of his former glory — the embodiment of the failure of imagination. His is the once-feared brave new world of science fiction writers. The future arrived much too quickly, while we were busy dismantling the past.

Ortega's question "who rules in the world?" has never seem more pertinent and vehement. Today that question is answered in the same disingenuous manner which it is posed: as an ideological begging of the question, which proves to be not a question at all but an ideological program. The categories that this insincere and fallacious pseudo thought aims to solidify are all too conspicuously bereft of any vital interior. But what else can vacuous thoughts aspire to? The impertinent, dead-end worldview that positivism has made the dominant — and often sole — of human reality comes about as the result of a morally and spiritually bankrupt philosophical materialism that does not possess the intellectual integrity to double check its premises. Thus, human reality today resembles Ortega's description in *Meditations on Quixote* — a forest that lacks trees.

How else to notice the forest, if not through an honest proximity to ourselves — an interior world that possesses the essential qualities to resist matter in all its tempting and destructive manifestations? Ortega's question as to who rules in the world? becomes ever more obvious now that Narcissus' chosen path of self-annihilation has become pandemic. Narcissus' likeness is no longer his but "theirs," "ours" — the contour of faceless anonymity — possessing the self-respect of a group, yet also liberating. The future beckons mass man to embellish himself in the new sensualism, in a centuries-old sought out liberation that promises an endless category of itinerant delights.

The Fight from Human Reality

T.S. Eliot contends that the essential characteristic of philosophy, that which makes it such an indispensable tool in deciphering reality, is the employment of common sense. Eliot's complaint that philosophy in the second half of the twentieth century "seems to provide a method of philosophizing without insight and wisdom" can only be grasped in its dire totality when we come to the frightening realization that he is speaking of "philosophers" and not just the man standing on the street corner. What then are we to conclude from Ortega's contention that the pathos of philosophers — in their current embodiment as "specialists" — is hardly different from that of non-philosophers?[182]

What T. S. Eliot seems to have in mind is not ignorance in the positive sense of the word: the understanding that one does not know. Eliot's fear is also Ortega's, that ignorance now means prescribed ignorance or the desire to proclaim that a lack of awe and wonder — anti-knowledge, let us call it — too, has a right to be considered part of the "hegemony" of "post-modern" knowledge.

Hence, the hope that Eliot retains in his measured and subdued criticism of the arid presupposition, if not the myopic and anemic output of logical positivism — a less ominous philosophy than today's nihilist "theory" — had to do with his reasoning that even blind alleys must occasionally be explored. He suggests: "certainly logical positivism is not a very nourishing diet for more than the small minority which has been conditioned to it." And then he goes on to add: "Yet, I believe that in the longer view, logical positivism will have proved of service by explorations of thought which we shall, in future, be unable to ignore; and even if some of its avenues turn out to be blind alleys, it is, after all, worth while exploring a blind alley, if only to discover that it is blind."[183]

This is indeed a fine hope, the kind of prescriptive moral/intellectual clarity that tries to forewarn that indiscretion always leads to painful dead ends. In the normal scheme of human reality this condition leads to a sobering up of the moral sense and of intellectual rigor; careful observation of the ways of small children or sincere seekers of truth easily bears out this truism. What, after all, is the cathartic impact of failure if no subsequent adjustment or careful calibration of the faculties that allow us to live in the world is made? Again, there is no space for "theoretical interpretation" over vital questions that have to do with our well-being — when we are

182. Josef Pieper. *Leisure: The Basis of Culture*, p. xii.

183. Ibid., p. xii.

the sole pilot of a small aircraft or hang glider, for instance. Once again, T.S. Eliot's notion of common sense comes into play.

However, the greater concern over our incessant, blind cry for revolt is that no spiritual, moral or social/political mechanism has been left in place that can serve as a self-corrective. The darkened alleys that Ortega's mass man explores are not only without end, but are also celebrated for their liberating qualities — a form of alleged vindication over "oppressive" common sense. The major contention that Ortega makes in *The Revolt of the Masses* is that revolt leaves no rock unturned and as such there cannot be a return to common sense, much less reason — correctives to corrosive and destructive anti-values — and it all takes place in the name of liberation. The nay-saying anti-valuation that Ortega identified in mass man — in the 1930s — is no less than what is venerated and cheered on today in the guise of "post-modernism." This phenomenon now embraces a new, "progressive" anti-valuation because traditionally such self-preservation has found it necessary to remake itself.

Again, what Pandora's box or the indiscretion of Hephaestus meant to the ancient Greeks can no longer have any meaning for us today. What can the idea of return mean to a time and age that posits no hierarchy of values as a part of the human condition?

Ortega's mass man — an invidious, destructive, social/political and opportunist force — can be envisioned today as the sanctioning body specializing in the intellectual lack of integrity and plagiarism that has leveled thought to mere "intertextuality." Mass man, Ortega tells us, can only triumph because his are values that are central to man's darker psyche and lower nature. Hence, once outfitted with the necessary intellectual, social/political instruments of direct action — namely "theory" —it can be said that a historical obstacle has been cleared for unfathomable and indefinite destruction. *The Revolt of the Masses* is very clear in this respect: history, human valuation and creation have always been ruled by the reality of noble and mass man. The purpose of civilization, Ortega argues, has always been to halt or hold back the tide of resentment, of hatred, incivility and man's raw irrational impulses. This is the genuine meaning, he tells us, of this ontological dichotomy. Who, then, are the forces behind mass man movements who cheer for "post-modernism" in all its nihilistic manifestations? Perhaps it is most appropriate to keep Ortega's importance of biography — knowing the Other from within — in mind in this respect?

Embracing "Post-Intelligibility," Contradictions and Incoherence

Ortega's suspicion that there can exist no genuine culture where there are no ideas, and that there are no ideas where there is no desire for truth on behalf of the thinker, serves as a strong indication of the progressive "liberation" of mass man. While he equates truth with resistance, difficulty, this is diametrically opposed to the modus operandi of mass man: an unqualified embracement of sensual and temporary "values." The questions that *The Revolt of the Masses* challenged us to entertain are rather daunting, even in 1930. While there existed a more genuine heterogeneous distribution of human values and personal perspectives in Ortega's time than in the initial decade of the twenty first century, mass man values exhibited the same homogeneous qualities that have always formed the backbone of this type of person. What is particularly interesting for us today is that with the spread and empowerment of mass man, we now encounter the kind of resentment that seeks to annihilate all aspects of heterogeneity of the human condition.

Ortega's notion of mass man is open-ended and fluid. The phenomenon that he identifies as mass man is not static. People become or embrace mass values through the choices and beliefs that it is in their power to embrace. Mass man's contrarian perspective exists today as just another form of life. This is why Ortega insists that mass man's scale of valuation is existential/ontological and not something that can be easily reduced to social/political abstractions of class, gender or race. As such, great violence has been done to sincere reflection on these eternal and universal questions by insisting in placating them under expedient ideological umbrellas.

Ortega's insistence that logic, reason and truth are no match for the violent and irrational irreverence of mass man is evident today when these tools have been attacked by self-contradictory and arbitrary conclusions. However, a key to understanding the impact of *The Revolt of the Masses* is its analytic acumen. The chapter on intellectuals, scientist, and other specialists clearly bears this out. In that chapter he makes the case that what was responsible for the acceleration of mass man's revolt is no less than the impetus that the former was given by "specialists" and intellectuals. Ortega concludes the chapter thus:

> But if the specialist is ignorant of the inner philosophy of the science he cultivates, he is much more radically ignorant of the historical conditions requisite for its continuation; that is to say: how society and the heart of man are to be organized in order that there may continue to be investigators.[184]

The reason that *The Revolt of the Masses* culminates with an analysis of the State and the question, Who rules the world? is highly suggestive of mass man finding its greatest impetus for action in the apparatus of the state. Barbarism, Ortega argues, is no longer an isolated historical phenomenon countervailing the trend of civilization. The periods of oscillation between these two forces, Ortega goes on to say, can no longer be accepted as accurate portrayals of the contemporary world. What has solidified the power of mass man over all forms of human existence is nothing less than the techniques of power that it now dominates. The politicization of culture, as perfected by State Communism and Nazism, Ortega demonstrates, is already a supreme example of the triumph of mass man.

In 1930, when Ortega published *The Revolt of the Masses*, he was able to draw empirical and current understanding from State Communism which was by then possessed by the idea of a State utopia — a

184. *The Revolt of the Masses*, p. 114.

scientific religion, in fact. He also paid close attention to the rising tide of Nazism and the pathos of its henchmen and women. What these two forms of positivistic socialism had in common was that both operated on the basis of an uncontained sophomoric resentment. This resentment, as we can easily verify today, was not established through the disorganized, wide-eyed whim of mass man. The techniques of social/political control, censorship and psychological terror which the aforementioned systems found indispensable for the effective execution of their pogroms were masterminded by materialist intellectuals — mass man, Ortega contends — but intellectuals none the less.[185]

The linking of technicism with triumphant philosophical materialism is Ortega's manner of demonstrating that mass man cannot exercise its most pent up resentment and destructive lower nature without the assistance of intellectual and "theoretical" foundation that vindicates its raw irrational primitivism in a semi coherent and sustainable manner. The significance of the question of who rules the world in the narrow scholastic and pedantic squabbles over Ortega's 1930s Spain cannot go unnoticed. All arguments that have taken this superficial route have failed in their failure to fully grasp the prescriptive nature of *The Revolt of the Masses*. In 1930, Ortega argued that the dehumanization of man that he witnessed would begin to demonstrate its true impact in thirty years.

Ortega argues — in no uncertain terms — that the State will serve in the future as the supreme form of barbarism by which mass man will exercise its raw might. "This is what State intervention leads to: the people are converted into fuel to feed the mere machine

185. Ortega discusses these themes in the latter part of the work, demonstrating that the value of this book, at least, has to do with his regard and respect for empirical data. For instance, in one passage, he writes: "So, in Moscow, there is a screen of European ideas — Marxism — thought out in Europe in view of European realities and problems. Behind it there is a people, not merely ethically distinct from the European, but what is much more important, of a different age to ours. A people still in process of fermentation; that is to say, a child-people. That Marxism should triumph in Russia, where there is no industry, would be the greatest contradiction that Marxism could undergo," p.137.

which is the State. The skeleton eats up the flesh around it. The scaffolding becomes the owner and tenant of the house."[186] Statism, in any of its varied forms can only defuse the vulgarity and dehumanization of social customs and cultural forms by appealing to the greatest common denominator: the appeasement of mass man. Ortega continues:

> Statism is the higher form taken by violence and direct action when these are set up as standards. Through and by the means of the state, the anonymous machine, the masses act for themselves.[187]

The "anonymous machine" is an apt description for resentment wherever it is cultivated. That the masses now — let us not forget that this term designates a self-annihilating, destructive pathos — works as a sanctioning body of rights, provisos and alleged entitlements without any recourse to reflection on the nature of duty. Ortega is no stranger to this argument when he writes:

> The nations of Europe have before them a period of great difficulties in their internal life, supremely arduous problems of laws, economics, and public order. Can we help feeling that under the rule of the masses the state will endeavor to crush the independence of the individual and the group, and thus definitely spoil the harvest of the future?[188]

A Surplus of Entertainment and Phantom Leisure

We end this inquiry as we began it, with an exploration into the essence of leisure. If we trace the trajectory of Ortega's early thought, from "Adam en el paraíso" ("Adam In Paradise"), *Meditations on Quixote*, *The Modern Theme* and *The Dehumanization of Art*, leading up to *The Revolt of the Masses*, only then can we build a true understanding of the theme and importance of the latter work.

"Adam en el paraíso" finds man alone, much like a proto first man. The themes that Ortega espouses in that work are those of an exis-

186. *The Revolt of the Masses*, p. 122.

187. Ibid., p. 123.

188. Ibid., p. 123.

tential nature. *Meditation on Quixote* is Ortega's best exposé of questions of ontology, consciousness and phenomenology. His portrayal of man in the forest, a symbol that designates human reality and the necessity of hearing our heart beating is a philosophical portrayal of Ortega's metaphor of "human existence as shipwreck" par excellence. What the solitary sojourner discovers in the forest is no less than himself, his life and the fullness, yet frightening reality of the human condition. The engrossing totality of the forest is equivalent to the cosmos. In the middle of the forest — there, amongst an endless array of objectifying entities — we encounter man, solitary and naked. Accustomed to always initially encountering trees, only then do we potentially make progress in our existentially inductive trek. Soon we begin to hear our own heart beating, Ortega suggests, as we encounter an entity that has no rival or equal, as the case may be, in our ability — Ortega refers to this as a calling — to come to terms with the notion that we never encounter the forest as a conglomeration of trees. Instead, what we manage to encounter there is nothing less than the self in the form of "I."

The Modern Theme continues in the development of these same themes, only there Ortega tackles the dangers of ideation, solipsism and the dead-end of becoming locked in unrealistic notions of consciousness. *The Dehumanization of Art* manages to capture the inquietude that is the essence of artistic creation.

Pretending to make sense of *The Revolt of the Masses* without first taking the time to make sense of the ontological/existential ramifications of this work will, for some readers, only prove to be a pretentious and arrogant show of ignorance. From such vacuous readings — and this is true today of all aspects of human existence, given the debilitating triumph of politicization for ideological expediency — sprout the seeds of unfounded suspicion and violence, both the hallmark of radical ideology. Ortega's presuppositions are simple and clear: Man is a morally free entity who must come to terms with the existential grounding of his own life.

Understood in its existential and phenomenological plenitude, *The Revolt of the Masses* is an exploration of man in society. While today most sociological accounts of man begin with the fallacious and arbitrary presupposition that man is merely a feather in the wind who's well-being is solely determined by an environmental behaviorism, Ortega, on the other hand, founds his reflection on man's existential freedom. And while many "models," "theories" and "debates" have attempted to foment the materialist anti-humanistic position of man as a carapace for positivists to dissect; it still goes without saying that man's primal freedom is not a subject for debate. As a fact, human freedom is further vindicated the more that arbitrary arguments against it are conjured up. The negation of human freedom is one of those ironies of human existence that are rife with contradictory and self-annihilating intent.

The silence and tranquility that forms such an integral part of *Meditations on Quixote* is completely ruptured, Ortega suggests, by the politicization of culture and every other facet of the human condition that he explores in *The Revolt of the Masses.* He makes this very clear in *The Dehumanization of Art*, a book that predates *The Revolt of the Masses* by five years. There he writes:

> Few people at this hour — and I refer to the time before the breaking out of this most grim war, which is coming to birth so strangely, as if it did not want to be born — few, I say, these days still enjoy that tranquility which permits one to choose the truth, to abstract oneself in reflection. All the world is in tumult, is beside itself, and when man is beside himself he loses his most essential attribute: the possibility of meditating, or with-drawing into himself to come to terms with himself and define what it is that he believes and what he does not believe; what he truly esteems and what he truly detests. Being beside himself bemuses him, blinds him, forces him to act mechanically in a frenetic somnambulism.[189]

It is also important to recall that the aforementioned notion of man being beside itself comes from the Spanish word *alteración* (to be beside oneself). To be beside oneself, Ortega argues, is much

189. *The Dehumanization of Art*, p. 178.

more than a psychological slogan of social disassociation or one that entails the social/political notion of "disenfranchisement" or even Sartre's idea of alienation. These are terms that because of their catch-phrase, bandwagon potential for expediency have come to lose their original fluidity and thus their ability to signify anything substantial about the underlying conditions of human existence. *The Revolt of the Masses* signals the start of an age of hyperactivity and hysteria.

The politicization of culture has made it impossible for the practice of the measured life of leisure that Pieper argues for. Leisure is not idleness. On the contrary, leisure means the ability to exist and remain alive within ourselves. *Alteración*, as Ortega refers to this form of inauthentic existence, is not a condition spanning from work. It is, however, a condition of a moral/spiritual emptiness. That the greatest concern of "man in revolt" should be the perpetual preoccupation with things, events, occurrences — in short, with external reality — bespeaks clearly and coherently of just what "revolt" has come to mean.

The reflective and vitally felt life creates an aura of influence around it that because of its self-awareness and understanding cannot easily fall prey to destructive external forces. The life of "revolt" must of necessity — at least as we have come to witness the progression of this empty term — rejects all genuine convictions, instead opting for mass appeal, immediate satisfaction and the allure of power that the politicization of all aspects of life entails. The true cause and subsequent self-annihilating source of "revolt" is the refusal to appropriate ourselves and our lives in the total scheme of reality that is the signature of the human condition. Owen Barfield makes this point explicitly:

> For instance, a non-participating consciousness cannot avoid distinguishing abruptly between the concept of "Man," or "mankind," or "men in general," on the one hand and that of "a man" — an individual human spirit — on the other.[190]

190. Owen Barfield. *Saving the Appearances: A Study in Idolatry*. Hanover: Wesleyan University Press, 1988, p. 183.

The Revolt of the Masses makes the point that when personal integrity is missing, it is then necessary to take to the cynical stance that attacks all hierarchical scale of values, and in the absence of moral and spiritual convictions what is valued most then is the "shared," communal, haphazard whim of a forged and false new "morality." Revolt will naturally gravitate toward a new world that is moved by the disingenuous clamor of everyman's liberation. When a cavernous moral, intellectual and spiritual cavity has been created where man formerly dwelled, what remains is the politicization of culture and human life. Revolt cannot help but to enjoy the fruits of "liberation" given that the incessant resentment of the hollow man fails to realize when it has indeed triumphed.

Glossary of Terms in Ortega's The Revolt of the Masses

1. Revolt of the Masses; *Rebelión de las masas* — The term "mass" alone is rarely use by Ortega. Instead, he employs it to mean mass-man: a kind of person who lives for the moment, vaguely cognizant of the world outside himself, having no particular vision for life and strongly opposed to others who do. "Revolt" signals a nihilistic rebellion against traditional values, norms and customs. However, revolt does not entail political revolt. On the contrary, for him revolt signifies an existential/moral crisis that is rooted in man's world-weariness — the death of wonder and awe. Revolt takes place, ironically, when human existence has become "secure," when man's surfeit of vital energy is no longer geared toward existential/moral salvation.

2. Agglomeration; *Aglomeración* — In itself, this term only suggests a geo-demographical condition. Ortega traces the phenomenon of agglomeration back to ancient history. The problem of agglomeration, however, is felt when the individuals who make

up large groups — a multitude — possesses the moral qualities of mass-man.

3. Minority; *Minoria(s)* — This term is synonymous with noble man. Minority, for Ortega has nothing to do with ethnic minorities. This word is taken to mean persons that resist the temptation and thus objectification of mass society and the values of mass-man.

4. Mass man; *Hombre-masa* — Ortega defines "mass man" throughout *The Revolt of the Masses*. However, any succinct definition of this term must include the following characteristics: a moral and qualitative essential human quality that is undifferentiated from other people. Mass man does not "impose any effort toward perfection" on himself. Also important is the understanding that mass man does not pertain to any given social/political/economic class.

5. Noble man; *Hombre noble* — Nobility for Ortega designates a stoic/existential engagement with human existence that springs from the recognition of the former as resistance. Nobility is a central quality of what Ortega also refers to as "minority man." However, nobility in this sense does not mean social or landed nobility.

6. Society; *Sociedad* — Society is made up of mass and minority man. Ortega argues that individuals have always existed in the multitude as political entities. Noble man possesses a greater degree of moral quality and otherwise, whereas the masses represent an assemblage, a "quantitative determination."

7. Quantity; *Cantidad* — Quantity is an existential/moral component in Ortega's work precisely because terms like "mass-man" and " revolt" are not construed as social/political in their respective meaning. Quantity (mass-man) is the opposite of quality (noble individualism).

8. Quality; *Calidad* — Quality is a moral/existential term that is characteristic of noble man. Because the terms "rebellion," "masses," and "social power" do not have a meaning that is primarily political, Ortega's analysis of human differentiation is commiserate with moral quality.

9. Hyperdemocracy; *Hiperdemocracia* — Ideally democracy should, Ortega argues, be ruled by a healthy respect for meritocracy. Hyperdemocracy is the opposite of a meritocracy. It is a state of democracy where the rule of law has been taken over by the unqualified worldview of mass man.

10. Proclamation of the Rights of Man; *Proclamación de los derechos del Hombre* — According to Ortega, the declaration of these inalienable rights should make man, "feel himself master, lord, and ruler of himself and of his life." The idea of individual rights entails personal autonomy. However, Ortega contends that this has not been the case, and that the notion of duty that these rights represent has become a burden and moral weight to modern man.

11. The Height of the Times; *La altura de los tiempos* — This term means "the tempo at which things move at present, the force and energy with which everything is done." This condition involves technology as well as the new morality that Ortega views as nihilistic. The height of the times also includes cultural, literary and aesthetic forms. This term signals a clash with tradition.

12. Modern; *Moderno* — Ortega uses this word in various ways. He defines it as:

> "The primary meaning of the words 'modern,' 'modernity,' with which recent times have baptized themselves, brings out very sharply that feeling of 'the height of time' that I am at present analyzing. *Modern* is what is 'in fashion,' that is to say, the

> new fashion or modification which has arisen over against the old traditional fashions used in the past. The word 'modern' then expresses a consciousness of a new life, superior to the old one, and at the same time an imperative call to be at the height of one's time. For the 'modern' man, not to be 'modern' means to fall below the historic level."

13. Decadence; *Decadencia* — Decadence, Ortega explains, suggests decline from a higher to a lower state. Decadence is opposed to standards of comparison. "There is only one absolute decadence; it consists in a lowering of vitality, and that only exists when it is felt as such."

14. Vital time; *Tiempo vital* — This term does not mean chronological, historical or cosmic time, but rather what Ortega refers to as mortal time. The relationship between vital, mortal time) and chronological time is an existential condition that man attempts to objectify in terms of chronological (shared), collective time. Implicit in this collective notion of time is the illusory belief that the fleeting quality of vital life can be assuaged the by social/political. This is an essential condition of revolt.

15. Barbarism; *Barbarismo* — Ortega defines barbarism as the absence of all standards, as these, "are the principles on which culture rests." The lack of a guiding principle creates in the mass man the unfounded belief that this is precisely how human history has always operated. However, Ortega's greatest concern in this respect is his notion that today, when it is commonly taught that everything is possible, decadence and barbarism too become viable options for man.

16. Life; *Vida* — While not exclusively a category found in *The Revolt of the Masses*, life in this work means vital life. It also means conscious possibility that is made known through the differenti-

ated concreteness that is our circumstance. Circumstance includes all of the ways in which life becomes conscious of itself, thus, vital life or what amounts to human existence.

17. Plant man; La *planta "hombre"* — This seemingly comic term is used by Ortega as synonymous with a materialist rendition of human existence, where the meaning of history is seen as a vast laboratory to experiment on man as a quantifiable (positivist), biological phenomenon. The significance of this term is that it negates the possibility of man's vital, inner quality.

18. Scientific Spirit; *El espíritu de la ciencia* — This idea illustrates Ortega's great regard for civilization and its historical trajectory. Thus, this term is not only relevant to scientific development, but rather to man's overall achievements and the pathos that motivates human creation. Ortega: "For, in fact, the common man, finding himself in a world so excellent, technically and socially, believes that it has been produced by nature, and never thinks of the personal efforts of highly-endowed individuals which the creation of this new world presupposed."

19. Nihilism; *Nihilismo* — This is a central theme that distinguishes mass from noble man. While the former embraces nihilism, the latter is appalled by it and guides his life as to transcend its grasp. Nihilism, Ortega argues, serves as a historical indicator that allows us to "anticipate the general structure of the future." Nihilism for Ortega cannot be separated from cynicism. The destructive impulse of nihilism is found in the cynic's, devils-advocate attitude, where what is argued for at an "intellectual" level, he is not willing to defend at a practical and moral one.

20. Inertia; *Inercia* — What has traditionally been seen as the domain of physical science, Ortega now turns into the anti-thesis of

an existential category. This is a condition of inertia or lack of the capacity to become awed by human accomplishment, and a form of volitional sterility that does not embrace reality as resistance. Inertia is what mass man feels in finding a world ready made.

21. Intellectual mass man; *Intelectual masa* — Ortega uses the phrase "the barbarism of specialization" to describe the self-satisfied arrogance of the specialist and the pride this type of person takes in the "narrow territory specially cultivated by himself." He defines the intellectual mass man as possessing no sense of overriding awe and wonder. But most importantly, Ortega calls the intellectual mass man a "learned ignoramus" because the only criteria that separates this type from overall ignorance is one of degree and not of kind. Also, the intellectual mass man is the direct opposite of what Ortega refers to as cultured people. Intellectual mass man lacks the central quality of the noble man: the desire to transcend itself. Intellectual mass man is solely moved by the career expediency of make-work research.

22. Vulgarity; *Vulgaridad* — Vulgarity is the fundamental and staple characteristic of mass man and, as such it is the antithesis of nobility. But vulgarity for Ortega denotes much more than moral values. Ortega conceives vulgarity instead to be a qualitative essence that attracts some people while repulsing others. The question of vulgarity in regard to mass man is not that the vulgar thinks itself of a higher moral quality and thus not vulgar at all, but that the vulgar proclaims itself to possess the right to imposes vulgarity as a right.

23. Culture; *Cultura* — Ortega uses this word to signify standards of appeal in matters of politics, aesthetics, intellectual disputes and economics. He ties culture to objective standards, and "standards as the founding principles on which culture rests."

24. Reason of unreason; *Razon de la no razon* — This is a term that signals the desire of the masses to impose vulgarity and irrational "notions born in the café" on society. One of the characteristics of "unreason" is the destruction of standards either out of resentment or vulgarity. Because the unifying quality of standards has been destroyed there remains no other models for conduct, aesthetic or otherwise. Now "everybody is the mass alone."

25. Technicism; *Tecnicismo* — Technicism is a term that Ortega uses to designate applied science. While science, that is, pure reason is responsible for creating the inventions of applied science; these two terms are not identical. Ortega argues that science is always linked with culture and as such it must be appreciated on its own merits, not necessarily for the goods that it produces.

26. The Self-Satisfied Man; *El hombre satisfecho* — The self-satisfied man is another characteristic of mass man. This term describes what Ortega suggests is a form of moral/spiritual fatigue that bargains for the enjoyment of the fruits of culture and civilization without recourse to their origin. This leads to a devil-may-care attitude that refuses to be moved by few conditions that pertain to the aforementioned categories. The self-satisfied person does not easily tolerate the idea of human life as resistance. Ortega: "He will intervene in all matters, imposing his own vulgar views without respect or regard for others, without limit or reserve, that is to say, in accordance with a system of 'direct action.'"

27. The Psychology of the Spoiled Child ; *La sicología del niño mimado* — This is another characteristic of mass man that explains his parasitic behavior as pertaining to the notion of entitlement associated with heirs. Ortega defines this aspect of mass man as: "This type which at present is to be found everywhere, and everywhere imposes his own spiritual barbarism, is, in fact, the spoiled child of

human history. In this case the inheritance is civilization — with its conveniences, its security; in a word, with all its advantages."

28. The Barbarism of Specialization; *La barbarie del especialismo* — Ortega uses this term to describe the negative aspects of scientific fragmentation of the history of ideas. This term can be seen as the opposite of a unified or "Renaissance man" perspective of knowledge. The major problem, as Ortega describes it, is not that, "the specialist 'knows' very well his own tiny corner of the universe; he is radically ignorant of all the rest." But rather that, "he proclaims it as a virtue that he takes no cognizance of what lies outside the narrow territory specially cultivated by himself, and gives the name 'dilettantism' to any curiosity for the general scheme of knowledge."

29. The State; *El Estado* — Even though Ortega offers a historical rendition of the development of the State, his analysis is mostly isolated to the role of mass man in its development. A characteristic of the State, Ortega argues, is that it is anti-individual, robbing man of his vitality. The State is also responsible for the "bureaucratization of life." Ortega credits the State with the destruction of "spontaneous social-action." He describes the origin of the State as follows: "The State begins when groups naturally divided find themselves obliged to live in common. This obligation is not of brute force, but implies an impelling purpose, a common task that is set before the dispersed groups. Before all the State is a plan of action and a program of collaboration. The men are called upon so that together they may do something. The State is neither consanguinity, nor linguistic unity, nor territorial unity, nor proximity of habitation. It is nothing material, inert, fixed, limited. It is pure dynamism — the will to do something in common — and thanks to this the idea of the State is bounded by no physical limits."

30. Demoralization; *Desmoralización* — Ortega attributes demoralization to the human desire to break free of all commands and moral restraints. However, this passion leads to what Ortega calls an "unemployed existence." This is also a characteristic of mass man that negates all forms of strife. Ortega writes: "By dint of feeling itself free, exempt from restrictions, it feels itself empty. An 'unemployed' existence is a worse negation of life than death itself because to live means to have something definite to do — a mission to fulfill — and in the measure in which we avoid setting our life to something, we make it empty."

31. Youth Culture; *Cultura de juventud* — Youth denotes for Ortega a time of more rights than obligations. This term is used as a symbol of a sophomoric stagnation of the moral sense. He writes: "The youth, as such, has always been considered exempt from *doing* or *having done* actions of importance. He has always lived on credit. It was a sort of false right, half ironic, half affectionate, which the no-longer young concede to their juniors. But the astounding thing at present is that these take it as an effective right precisely in order to claim for themselves all those other rights which only belong to the man who has already done something."

32. World; Mundo — World is synonymous with our individual circumstances. World, as circumstance, signifies mans "sum-total of vital possibilities." The world is a sort of theater where we play out our existence. I am in the world because this is what human existence entails: an entity capable of self-consciousness that is always in a given situation. Grasped in phenomenological terms, the world acts as the horizon of my life.

Bibliography

Primary Sources

Ortega y Gasset, José. *Meditaciones del Quixote*. Madrid: Cátedra, 1984.

_________. *Ideas y Creencias*. Madrid: Espasa-Calpe, S.A., 1976.

_________. *Caracteres y Circunstancias*. Madrid: Afrodisio Aguado, S.A. — Editores-Libreros, 1957.

_________. *La redención de las provincias: Escritos políticos, II (1918/1928)*. Madrid: Revista de Occidente, 1973.

________. *En torno a Galileo/El Hombre y la gente*. México: Editorial Porrua, S.A., 1985.

________. *El Espectador, II*. Madrid: Revista de Occidente, 1969.

________. *Papeles sobre Velásquez y Goya*. Madrid: Revista de Occidente, 1987.

________. *Meditación de la Técnica*. Madrid: Espasa-Calpe, S.A. 1965.

________. *La deshumanización del arte y otros ensayos de estética*. Madrid: Espasa-Calpe, S.A., 1987.

________. *Kant, Hegel, Scheler*. Madrid: Revista de Occidente, 1983.

Works by Ortega in English Translation

Ortega y Gasset, José. *The Revolt of the Masses*. New York: W.W. Norton & Company, 1960.

_________. *An Interpretation of Universal History*. New York: W.W. Norton, 1975.

_________. *Meditations on Quixote*. New York: W.W. Norton & Company, 1963.

_________. *The Modern Theme*. New York: Harper & Row, Publishers, 1961.

_________. *What is Philosophy?* New York: W.W. Norton & Company, 1964.

_________. *History as a System and Other Essays Toward a Philosophy of History*. New York: W.W. Norton & Company, 1962.

_________. *Man and People*. New York: W.W. Norton & Company, 1963.

_________. *Meditations on Hunting*. New York: Charles Scribner's Sons, 1982.

_________. *The Origin of Philosophy*. New York: W.W. Norton & Company, Inc, 1967.

_________. *Man and Crisis*. New York: W.W. Norton & Company Inc. 1962.

Secondary Sources

Alba-Buffill, Elio. *Cubanos de dos siglos: Ensayistas y criticos*. Miami: Ediciones Universal, 1998.

Aron, Raymond. *The Opium of the Intellectuals*. New Brunswick: Transaction Publishers, 2004.

Barfield, Owen. *Saving the Appearances: A Study in Idolatry*. Hanover: Wesleyan University Press, 1988.

Barrett, William. *Time of Need: Forms of Imagination in the Twentieth Century*. New York: Harper & Row, 1972.

Barzun, Jacques. *The Culture We Deserve*. Hanover: Wesleyan University Press, 1989.

________. *From Dawn to Decadence: 500 Years of Western Cultural Life, 1500 to the Present.* New York: HarperCollins, 2000.

Baudrillard, Jean. *Cool Memories II.* Translated by Chris Turner. Durham: Duke University Press, 1996.

________. *The Ecstasy of Communication.* Translated by Bernard & Caroline Schutze. New York: Autonomedia, 1987.

Burke, Edmund. *A Vindication of Natural Society: Or, A View of the Miseries and Evils Arising to Mankind from Every Species of Artificial Society.* Edited by Frank N. Pagano. Indianapolis: Liberty Fund, 1982.

Cabrera Infante, Guillermo. *Mea Cuba.* Barcelona: Plaza & Janes Editores, 1992.

Camus, Albert. *The Rebel: An Essay on Man in Revolt.* Translated by Anthony Bower. New York: Vintage International, 1991.

________. *Between Hell and Reason: Essays from the Resistance Newspaper Combat, 1944-1947.* Translated by Alexandre de Gramont. Hanover: Wesleyan University Press, 1991.

________. *Lyrical and Critical Essays.* Edited by Philip Thody. Translated by Ellen Conroy Kennedy. New York: Vintage Books, 1970.

Canetti, Elias. *Crowds and Power.* Translated by Carol Stewart. New York: The Seabury Press, 1978.

________. *The Secret Heart of the Clock: Notes, Aphorisms, Fragments. 1973-1985.* New York: Farrar Straus Giroux, 1989.

Clark, David K. *Empirical Realism: Meaning and the Generative Foundation of Morality.* Lanham, Maryland: Lexington Books, 2004.

De Madariaga, Salvador. *Portrait of a Man Standing.* University: University of Alabama Press, 1968.

Edwards, Jorge. *Persona Non Grata.* Barcelona: Tusquets Editores, 2000.

Ferrater Mora, José. *Man at the Crossroads.* New York: Greenwood Press, Publishers, 1968.

Ferry, Luc and Alain Renaut. *French Philosophy of the Sixties: An Essay on Antihumanism.* Amherst: The University of Massachusetts Press, 1990.

________. *Heidegger and Modernity.* Translated by Franklin Philip. Chicago: The University of Chicago Press, 1990.

Fleming, Bruce. *Science and the Self: The Scale of Knowledge.* Dallas: University Press of America, 2004.

Forster, E. M. *Aspects of the Novel.* San Diego: Harcourt, Inc., 1955.

Fukuyama, Francis. *The End of History and the Last Man.* New York: The Free Press, 1992.

Gombrowicz, Witold. *A Guide to Philosophy in Six Hours and Fifteen Minutes.* New Haven: Yale University Press, 2004.

Gonzalez, Pedro Blas. *Human Existence as Radical Reality: Ortega y Gasset's Philosophy of Subjectivity.* St. Paul: Paragon House, 2005.

__________. *Fragments: Essays in Subjectivity, Individuality and Autonomy.* New York: Algora Publishing, 2005.

__________. "Half a Century of Philosophy." *Philosophy Today*, Summer 1998, Volume 42:2: 115-125.

Gray, Rockwell. *The Imperative of Modernity: An Intellectual Biography of José Ortega y Gasset.* Berkeley: University of California Press, 1989.

Guitton, Jean. *El Trabajo Intelectual.* Mexico: Editorial Porrus, S.A., 1994.

Hoffer, Eric. *The True Believer: Thoughts on the Nature of Mass Movements.* New York: Harper & Row, Publishers,

________. *In Our Time.* New York: Harper & Row, Publishers, 1976.

________. *First Things, Last Things.* New York: Harper & Row, Publishers, 1971.

________. *The Temper of Our Time.* New York: Harper & Row, Publishers, 1967.

Ingenieros, José. *El Hombre Mediocre.* Buenos Aires: Editorial Porrua, S.A., 1978.

________. *Las Fuerzas Morales*. Buenos Aires: Editorial Losada, 1968.

James, William. *The Selected Letters of William James.* Edited by Elizabeth Hardwick. New York: Doubleday, 1961.

Jaspers, Karl. *Way to Wisdom*. New Haven: Yale University Press, 1954.

Judt, Tony. "Review of Louis Althusser, The Paris Strangler." *New Republic*. March 7, 1994.

Lawson, Lewis and Victor A. Kramer. Editors. *Conversation with Walker Percy*. Jackson: University Press of Mississippi, 1991.

Lavelle, Louis. *The Dilemma of Narcissus*. Burdett: New York: Larson Publications, 1993.

________. *Evil and Suffering*. New York: The Macmillan Company, 1963.

Larrayo, Francisco and Edmundo Escobar. *Historia de las Doctrinas Filosoficas en Latinoamérica*. México, D.F., 1968.

Lefevre, Perry D. Editor. *The Prayers of Kierkegaard*. Chicago: The University of Chicago Press, 1963.

Levy Henri, Bernard. Editor. *What Good Are Intellectuals?* New York: Algora Publishing, 2000.

Llera Pinera, Humberto. *Panorama de la Filosofia Cubana*. Washington: Union Panamericana, 1960.

Lewis, C.S. *The Problem of Pain*. New York: Macmillan, 1973.

________. *The Abolition of Man: How Education Develops Man's Sense of Morality*. New York: Macmillan Publishing Co., 1978.

Lewis, Wyndham. *Time and Western Man*. Edited by Paul Edwards. Santa Rosa, California: Black Sparrow Press, 1993.

Marcel, Gabriel. *The Philosophy of Existentialism*. New York: Carol Publishing Group, 1995.

Marias, Julian. *José Ortega y Gasset: Circumstance and Vocation*. Norman: University of Oklahoma Press, 1970.

________. *Philosophy As Dramatic Theory*. University Park: Pennsylvania State University Press, 1971.

________. *Tratado Sobre la Convivienca: Concordia sin acuerdo*. Barcelona: Ediciones Martinez Roca, 2000.

Musil, Robert. *The Man Withouth Qualities.* Translated by Sophie Wilkins. New York: Vintage International, 1996.

Nietzsche, Friedrich. *On the Genealogy of Morals.* Translated by Walter Kaufmann and R. J. Hollingdale. New York: Vintage Books, 1969.

________. *Ecce Homo*. Translated by Walter Kaufmann. Vintage Books, 1969.

________. *Beyond Good and Evil*. Translated by Walter Kaufmann. New York: Vintage Books, 1966.

Oakeshott, Michael. *Rationalism in Politics and Other Essays*. Indianapolis: Liberty Press, 1991.

__________. *The Voice of Liberal Learning*. Indianapolis: Liberty Fund, 2001.

Paz, Octavio. *El Laberinto de la Soledad*. Mexico: Coleccion Popular, 1977.

Pieper, Josef. *Leisure: The Basis of Culture*. Indianapolis: Liberty Fund, 1999.

Putnam, Hilary. *The Many Faces of Realism: The Paul Carus Lectures*. LaSalle, Illinois: Open Court, 1991.

Quimette, Victor. "Ortega y Gasset and the Limits of Conservatism." José Ortega y *Gasset: Proceedings of the Espectador universal International Interdisciplinary Conference*, p. 65-71.

Renaut, Alain. *The Era of the Individual: A Contribution to a History of Subjectivity*. Princeton: Princeton University Press, 1999.

Revel, Jean-Francois. *On Proust*. New York : The Library Press, 1972.

Ribas, Armando P. *El Fin de la Idiotez y la Muerte del Hombre Nuevo*. Miami: Ediciones Universal, 2004.

Ripoll, Carlos. *La Generacion del 23 en Cuba*. New York: Las Americas Publishing Co., 1968

Rougemont, Denis de. *Love Declared : Essays on the Myths of Love.* Translated by Richard Howard. New York: Pantheon Books, p. 163.

Scruton, Roger. *Thinkers of the New Left.* London: Claridge Press, 1985.

Sertllanges, A.D. *La Vida Intelectual.* México: Editorial Porrua, S.A., 1994.

Skinner, B.F. *Beyond Freedom and Dignity.* New York: Bantam Books, 1990.

Simmel, Georg. *On Individuality and Social Forms: Selected Writings.* Chicago: The University of Chicago Press, 1971.

Sokal, Alan. *The Sokal Hoax: The Sham that Shook the Academy.* Edited by the editors of Lingua Franca. Lincoln: University of Nebraska, 2000.

Sowell, Thomas. *The Vision of the Anointed: Self-Congratulation as a Basis for Social Policy.* BasicBooks: New York, 1995.

Spengler, Oswald. *The Decline of the West.* New York: The Modern Library, 1962.

Sartre, Jean-Paul. *Situations.* Translated by Benita Eisler. Greenwich, Connecticut: Fawcett Publications, Inc., 1966.

_______. *El Diablo y Dios.* Buenos Aires: Alianza Losada, 1981.

Stegner, Wallace. *Crossing to Safety.* New York: The Modern Library, 2002.

Templeton, Kenneth S., Jr., Editor. *The Politicization of Society.* Indianapolis: Liberty Fund, 1979.

Toffler, Alvin and Heidi. *Creating a New Civilization: The Politics of the Third Wave.* Atlanta: Turner Publishing, Inc., 1995.

Unamuno, Miguel. *Mist: A Tragicomic Novel.* Translated by Warner Fite. Urbana: University of Illinois Press, 2000.

_______. *Del Sentimiento Tragico de la Vida En Los Hombres y En Los Pueblos.* Madrid: Editorial Plenitud, 1966.

_______. *Un Siglo de Ortega y Gasset.* Madrid : Editorial Mezquita, 1984.

_______. *Selected Works of Miguel de Unamuno: The Private World.* Vol. 2. Princeton: Princeton University Press, 1984.

Vargas Llosa, Mario. "La rebellion de las masas." *Diario Las Americas.* December 4, 2005.

Vega, David Rodriguez. Editor. *Palabras de Ortega en Chile. Santiago*: Centro Cultural de España en Santiago de Chile, 2005.

Verdes-Leroux, Jeannine. *Deconstructing Pierre Bourdieu: Against Sociological Terrorism from the Left.* New York: Algora Publishing, 2001.

Vitier, Medardo. *La leccion de Varona.* México, D.F.: Jornadas, 1945.

Wagner, Geoffrey. *Wyndham Lewis: A Portrait of the Artist as the Enemy.* New Haven: Yale University Press, 1957.

Weaver, Richard M. *In Defense of Tradition: Collected Shorter Writings of Richard M. Weaver, 1929-1963.* Indianapolis: Liberty Fund, 2000.

Whitehouse, Harvey and Robert N. McCavley. Editors. *Mind and Religion: Psychological and Cognitive Foundations of Religiosity.* Walnut Creek, California: AltaMira Press, 2005.

Wood, Peter. *Diversity: The Invention of a Concept.* San Francisco: Encounter Books, 2003.

INDEX